In a Patch of Fireweed

In a Patch of Fireweed

Bernd Heinrich

Harvard University Press
Cambridge, Massachusetts, and London, England

First Harvard University Press paperback edition, 1991

Library of Congress Cataloging in Publication Data
Heinrich, Bernd, 1940 –
   In a patch of fireweed.

   1. Heinrich, Bernd, 1940 –   .2. Insects.
3. Entomologists — United States — Biography.   I. Title.
QL31.H42A33   1984   574   83-13043
ISBN 0-674-44548-1 (alk. paper) (cloth)
ISBN 0-674-44551-1 (paper)

To my parents
and to my mentors — Phil, Dick, and Bart

Preface

I am a biologist, and research is a major part of my life. It can be tedious and difficult, but on the whole it is full of adventure and I love it. Often I have been frustrated with the journal articles that come out of the research because only the finished results are given. All the excitement of the process has been squeezed out so that the results will conform to certain expected standards necessary for clear and objective scientific communication. (I expect no less from other researchers who communicate to me through the journals.)

This book is not really for scholars, then. I am more concerned about conveying to a general audience what motivates someone to get into natural science. My hope is to capture here some of the feeling of science that I have had to leave out of my other writings: the sounds and sights, the endless chores and happy accidents, the obsessions, the wonder of it all.

The first three chapters describe personal experiences that helped to shape my later perceptions and work. My intention was not to write autobiography but to tell about the natural links forged between one's life and a life in science. Other chapters, on ant lions, wasps, and moths, provide results that will be written up for scientific journals. Finally, in the last chapter I talk about some possible directions for future research and how my involvement with biology has shaped my view of life.

The manuscript evolved not only from research but also from observations, thoughts, and ideas written down over the course of many years. I owe special thanks to William Patrick of Harvard University Press for encouraging me to try to combine such eclectic material into a single manuscript. Joyce Backman gave invaluable help to see it through. I am also grateful to my sister Ursula (Ulla) Wartowski and to my parents, Hildegard and Gerd Heinrich, for refreshing my memory with details I had forgotten or suppressed. My wife, Margaret, gave indispensable assistance in the field, encouragement, support, and always helpful criticism. Bert Hölldobler, Bill Jordan, Ernst Mayr, and David Stanley-Samuelson offered many useful suggestions on earlier versions of the manuscript. I also thank Ann Fortner, who patiently and accurately typed so many drafts.

Drawing is a way for me to get reacquainted with my feelings for the subject, and I include some sketches here not as scientific illustrations but as fond recollections.

Contents

In a Patch of Fireweed

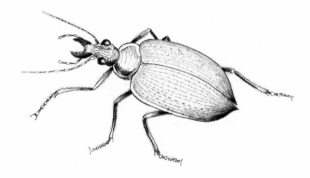

Flight into the Forest

I remember the fresh green grass waving in the slight breeze under a brilliant blue sky, and in this grass stood a small bush with pale brown stems. On the stems were bright yellow flowers more beautiful than any I'd ever seen before. My urge to pick them was all-absorbing, and I forgot the black plume of smoke of an airplane falling from the skies, and I hardly heard the loud wailing of the sirens that slowly rose in pitch and then dropped, nor did I pay attention to the dull muffled thunder in the distance.

As I was reaching for the flowers, Mamusha yanked on my arm. "Komm, schnell, schnell!" There was a strange, serious look in her eyes. In an instant we'd entered through the steel door into a dark cavern underground where people sat like mummies along the cement walls. The thunder could still be heard. These are early memories, but they connect me even now to the present. I do not like forsythia.

I can recall very little of Borowke, our large farm in what is now Poland. I was only four years old when we left. But from the album of photographs we took with us and from the constant reminiscing of my parents, it must have been heaven on earth. There were wooded hills, ponds, marshes, and rolling meadows with horses, cows, and sheep.

We had been landed gentry. My grandfather was a successful physician in Berlin, my grandmother an accomplished painter who had studied in Paris. My father was an amateur biologist who had spent several years on expeditions to the wilds of Burma, Celebes, and Halmahera to collect rare birds for the New York, Chicago, and Berlin natural history museums. Visitors from abroad came often to Borowke to enjoy the country air and to stay for as long as they liked in our large house surrounded by massive chestnut trees planted by my great-grand-father.

Then the dark cloud of war enveloped us all. The bucolic serenity, and our security, vanished like the sun, and we came to live at the edge of hunger. In the immediate sense this happened because we were caught between Communists, Nazis, and British and American bombers. But, as much as anything else, somewhere along the line it also came to be because of misunderstanding and mistaken ideas about the application of biological theories on the struggle for existence and the "natural" order of things.

To the Communists at that time, the superficial order of the anthill was natural, and to achieve that specific automatic order through obedience they aimed to kill off people like us, the so-called intelligentsia, who were thought to have nothing to contribute. For their part, the Nazis turned to the highest scientific and philosophical authorities to sanctify their ideology. They leaned on Darwin and Nietzsche, like drunks leaning on lampposts, not for illumination but for support. The Nazis picked one specific human type out of the innumerable ones that nature had evolved through natural selection, and they proclaimed that only this **one** was "fit" and entitled to life. We were worried, because my grandmother was Jewish. Papa revived some old contacts from his time as a pilot in World War I, and they helped him to enlist in the Luftwaffe. We hoped this would keep the Nazis from ferreting us out. It did.

In hindsight, it seems that the difference between the life and death of millions hinged ultimately on an ignorance of biology or on distorted application of these ideas on how things ought to be. At that time, back in 1944, of course I thought none of this. I only reacted to immediate circumstances affecting my feelings of well-being. My parents thought more into the future, worrying about survival under conditions imposed on us by forces beyond our control. It was too late then to change circumstances, and one could only hope to live through them. Our next biggest fear was the Russian army.

We had been constantly reassured that there was no reason to fear a Russian invasion. But suddenly, one night, it was suggested that we evacuate. Many landowners stayed, believing that, since they had done no harm, no harm would come to them. It was a wrong assumption and, like many assumptions based only on hope and feelings, it would prove deadly to them.

I was excited to be going on a big trip, although I would miss George, a British prisoner of war who worked on our farm. He had helped my older sister with her English, and he told me about the gnomes that he said lived in the forest. Before we left he had taken me into a birch-alder bog, showing me little hillocks of moss they presumably hollowed out for living quarters. I had never seen any gnomes, but George assured me that they always hid in the daytime and would only dance about at night in the moonlight on top of a big mushroom. They were obviously not only small, unobtrusive, and mysterious, but also very clever to live in such total secrecy. I wondered if I would ever see them now that we were leaving Borowke.

It was snowing when we left. I was bundled up snugly against the cold, smiling and saying Auf Wiedersehen to farmhands who stayed. They seemed to hug a little harder than usual and to look into my eyes a few seconds longer. Then we left — into the night riding on a wagon with only some spare clothing, a little food, a clock of my great-grandfather's, and the album of photographs. The wheat fields, the gentle hills, my grandmother's paintings of flowers and butterflies, the horses and cows, the forests, all were left behind.

Leaving one's land is not an easy thing to do. The chaffinches who nested in the chestnut trees by our house escaped the winter by flying south for thousands of miles, but they were drawn north again in the spring to that tiny pinpoint on the map they knew so well, to their home. We knew we'd never return. We'd have to put down roots elsewhere, but that goal was not very high on our list of priorities right then.

Mamusha, my two sisters Marianne and Ulla, and I did not leave Borowke until we heard artillery and knew that the fighting was getting close. Our immediate goal was to try to reach an airbase in the north where Papa was stationed.

We made very slow progress. The roads were already jammed full of sleighs, wagons, horses, and endless streams of people carrying a few prized belongings, utensils, and food. The German army was retreating. Everyone was scouring the land for food like a hungry plague of migrating locusts. After a while we could hardly move forward at all on the roads clogged with fleeing humanity. Mamusha and Ulla (my older sister) were increasingly concerned about how to negotiate through this bedlam to make it north. We moved off the road and decided to hide in a farmhouse in a remote village in order to see what might develop. Mamusha and Ulla watched the highway, talked to the refugees, and gathered information about the Russians who were coming on as steadily as phalanxes of army ants marching in the jungle. A retreating Panzer division from Kurland was there, and the officers told us that it was impossible for us to remain, for the women, if caught by the Russians, would be raped and killed. They offered to take us with them, and we accepted. We rode for several weeks in a tank used for medical supplies, while the soldiers continued to fight and to hold off the fast-advancing Russians.

One evening we moved with the Kurlanders and their tanks into a deserted village. The soldiers slaughtered some pigs and started to cook them over open pits. Everyone had a chance to wash with warm water. That was already a big treat and, having been hungry for so long, we were all looking forward to having a good Schweinebraten as well. Then, as the aroma of roast pork saturated the air, an alarm sounded. The soldiers dashed for their tanks, and nobody stayed to enjoy the pork. Later we learned that, while we were leaving the village at one end, the Russians were moving into it at the other end, barely a kilometer away. It was they who undoubtedly enjoyed the Schweinebraten.

After this and other close encounters, our adventures with the Panzer unit ended. They had to stay and fight. Civilians were in the way, and we would have to make it by ourselves. They had only one option — at least we still had the freedom of trying to beat the odds. So again we struck out on our own. Eventually we managed to reach Papa's base. But when we got there, he was gone.

He had waited. Steady streams of people had flowed north on the highways for days, passing the airbase. It had become clear to him then that the Russians were advancing rapidly, pushing the civilian population ahead of them. We did not come, and there was no word from us. Having expected to see us long ago, he had traveled in desperation to Berlin to our alternate place for contacts. Fortunately, however, we waited and eventually he did come back to find us.

Where should we go next, and how? Because Mamusha and Ulla had befriended the base commandant, Papa was given permission to escort us out of the danger zone by taking us, and some other refugees, on an old wood-burning truck that was being sent to the airbase in Stolp, to drop off antiaircraft equipment. But before we reached Stolp, we learned that the Russians had penetrated, cutting off all roads to the west. We were completely encircled, and the whole area was now aptly called "Der Kessel" (the Kettle). Escape by land was quite impossible, and the Russians were closing in from all sides. Might there still be a chance to escape by air?

A unit of the German army, including our friends with the Panzers, was continuing to fight east of us. The Junkers that flew supplies to these surrounded and doomed troops stopped in Stolp, and then again flew out of the Kettle. We could try to get into one of those planes, to get out — to anywhere; women and children had priority, aside from military cargo. But the airport was soon to be abandoned and blown up. When the shooting was getting so close that it could be heard in the distance, we managed to get into one of the last planes scheduled to leave. The tankmaster of the airport didn't want to give our pilot fuel — he wanted to save it for the few remaining fighter planes that had not yet been shot down. Papa (who was a teetotaler) bribed him with a bottle of liquor specifically carried for such emergencies. There is a market for mind-altering drugs in bad times, and we got our fuel. Then one of the *propellers broke as the engine was being started. Calls were made quickly to Berlin. An official answered that a propeller could be provided, but only if his girlfriend, who happened to be at Stolp, could also be taken out on the plane. It was a deal. The next day the propeller arrived.*

Next, one of the two engines wouldn't start. Papa talked the Junkers pilot into trying to fly with just one engine. As we were rolling down the runway, the plane swerved to one side. Something was wrong with the steering mechanism as well. That left us little choice. It was snowing hard as we left the plane and trudged toward the woods to await our fate with the Russian army. Just as we were leaving, the pilot called us back: "Komm schnell—Ich hab's gefunden—Ich hab's gefunden!" (Come quick, I have found it, I have found it!). The pilot had found a minor cable malfunction that he could fix and we took off, just in the nick of time, too, because the demolition squad was ready to ignite the bombs that were already set to destroy the airstrip. Apparently the bribed tankmaster had provided only the bare minimum of fuel, and our pilot wanted to land on an airfield halfway to our destination, hoping to get more fuel. He and Papa talked it over and decided we might have just barely enough after all. It was a good decision because, when we flew over the airport where we were thinking of landing, we saw black puffs of exploding flack — they were shooting at us! The airport was already in the hands of the Russians.

We made it to the airport at Demine and stayed there a few days. We were trying to find a means to get to Schwerin, where Papa was to report for duty commanding the defense of the airbase there. He bribed the woman in charge of the train station, and she let us board a cattle car. After one day and one night we reached Schwerin. We had only a loaf of bread to eat on that train. It was very good bread, and we were not displeased. (Usually, only when reality does not meet expectations does unhappiness settle in.)

In Schwerin, Papa was instructed to give young boys—the new draftees—lessons in National Socialism. He protested that he had no lessons himself, so how could he instruct? The SS officer in charge of political orientation told him that this could be easily remedied and sent

him — even though the collapse of the Third Reich was obviously immi-
nent — to take a course at Oranienburg near Berlin. He was there with
other Luftwaffe people, and to a man they all had the same low opinion of
the matter. At the end of the course each man was required to give a
speech. Papa's was on his experience of catching rare birds in the jungles
of Celebes. He said it was a big hit with the officers.

While Papa was being indoctrinated, we stayed at the airport in
Schwerin, where there were almost daily bombing raids by American
B-17s. It was here, on the meadow close to the airport barracks, where I
saw the black plume of smoke and the forsythia that I didn't have time to
pick because we had to run to the bomb shelter.

After Papa returned, he was assigned to oversee some guard dogs and
some young recruits to defend the airport. The troopers scouted daily
and reported to him. Then, on May 2, we heard that Hitler had killed
himself, and the following morning the patrols told Papa that enemy
tanks were coming — they would be here in about half an hour. The
Russians and the Americans were thought to be equally close and we
didn't know which would arrive first. Papa quickly persuaded the
quartermaster to open up the supplies and to let everyone take what
they needed. Then he jumped on his bicycle and went to hide in the
woods. As usual, my parents had anticipated events, and it had been
worked out beforehand that Papa would hide under a certain pile of
brush in a clearing in the forest. Ulla and Mamusha were to bring him
food for several days, until the next step could be taken.

While Papa had been at the airport, we lived in the hayloft of a barn in
the nearby village of Sülzdorf. Now, when he was hiding in the forest, we
could hear the artillery duels. The shooting was getting closer all the time.
We hid in the cellar of the farmer's house, wondering who would get here
first, the Americans or the Russians. Suddenly the guns stopped; the
silence was shattering. Ulla, consumed by curiosity, sneaked into the
barn to look out a tiny window onto the street. What she saw almost took
her breath away: a tank with a big white star painted on it. She knew the
insignia of the Russians was a star, and she was frightened. Trying to
figure out where to hide next, she heard in English, "Hey, come out,

come out, you bastards!" She said later that no words had ever sounded sweeter to her ears. (Ulla had learned English in high school and George had helped a little to enlarge her vocabulary, but she had not learned that the color of the insignia made all the difference.) She raced back yelling "Die Amerikaner sind hier — die Amerikaner sind hier!" and then ran out, practicing her English, "Welcome to Sülzdorf!"

Meanwhile Papa was lying on the damp earth under the brush when a great commotion broke out — artillery fire, infantrymen running by, and tanks rumbling past, crashing over the fir trees. Then it was quiet again. Later other men arrived, speaking German. It was apparent that these were SS men, the dread internal police of the Nazis, who were probably getting ready to surrender as a group or to metamorphose into civilians by changing clothes. They would likely shoot Papa if they found him because they would not want witnesses. On the other hand, if they were following Hitler's orders and not surrendering, they would certainly shoot him as a deserter. The SS men started a fire, taking wood from the very pile under which Papa lay. But their lunch break was cut short when they heard the sound of tanks, and they felt compelled to disappear in some hurry. In their great rush they left behind many supplies, including a large crate of butter, a typewriter, and a big box full of Iron Crosses (military decorations). These later turned out to be excellent items of barter for food with the occupying Americans, and the typewriter ultimately came with us to America.

The German army was disbanded, and the countryside was strewn with supplies. Horses ran loose, and Ulla captured three of them. She also found saddles and harnesses. Ulla, who had always loved riding in Borowke, was in her element. She even gave the American GIs riding lessons, and they soon were calling her "Blondie." The GIs were from a unit that named themselves the Grasshoppers. One day a Grasshopper lieutenant came with a German car and several cans of gasoline to the hayloft where we were living. He said to Ulla, "Blondie, please take your family and move west. Don't stop till you cross the Kaiser Wilhelm Canal." "Why?" Ulla wanted to know. "Trust me," the Grasshopper lieutenant said, "it will be better for you." Ulla asked him why he should be concerned about us. He said, "Blondie, you're like my girl back home. I wish you well."

There could be no doubt that the man was sincere. Perhaps he knew something he couldn't tell us. We trusted him, but the problem was that neither Mamusha nor Ulla knew how to drive a car, and we still wanted to keep Papa in the background as much as possible. As chance would have it, we found a canvas-covered army supply wagon. We hitched on Ulla's horses, loaded the wagon with butter, the Iron Crosses, the typewriter, as well as Papa disguised as a grandfather (he was a balding forty-seven), and struck out toward the west as the Grasshopper lieutenant had advised.

All went well until we came to the Kaiser Wilhelm Canal. The crossing was guarded by the British. No civilians were to cross, by order of the British military commander. We didn't know it then, but this canal was to be the boundary between East and West Germany. To us it was simply the boundary between British and American territory; we never dreamed that this land would eventually be handed over to the Russians.

The British guarding the crossing had a first lieutenant from Poland among them, one of our countrymen; Ulla and my parents spoke fluent Polish. The lieutenant persuaded his British friend, a guard, to help us pass. We camped there for several days, waiting for the moment when this friend had guard duty before dawn. The illegal action went smoothly. We had followed the GI's advice, and we had crossed the Kaiser Wilhelm Canal. We had our lives and our wits, the worst was over. A new future lay ahead.

The times were not easy, though. The biggest problem was filling our bellies. Papa decided that the best chance of finding food would be in the forest. We came across a large reserve called the Hahnheide, and within it a small empty hut used before the war by a nature club from Hamburg. The forester in charge gave us permission to move in. The people in the neighboring villages had already taken so many refugees into their homes that there was little room for more.

We lived deep in the forest for five years. We had no work and hardly *ever any money. The civilization that seals most of us off from the stark reality of existence had broken down. We were totally immersed in nature. Like most animals, our major concern was with finding food. Especially I remember foraging for berries.*

It was late in the afternoon. I must have picked a million raspberries already. Ulla, Mamusha, and Marianne were picking too. We worked fast to clear out the key patch we had discovered, before any of the other refugees from the nearby towns might find it. It was boring to pick raspberries (and later in the summer blueberries), but occasionally I would find a big fat green caterpillar with red spots and tufts of short bristles and would take it back to our hut under the tall spruce trees by the brook. I would feed it fresh raspberry leaves every day, watch it spin a cocoon out of brown silk, and at long last see a beautiful silkmoth emerge. That was exciting and compensated somewhat for the tedious chore of picking berries all day long.

I didn't like picking berries because I had to move so slowly, from bush to bush. I much preferred picking mushrooms, when I could run at will through the damp forest, feeling the soft green moss under my bare feet. Every place in the forest was different. The orange Rehfüsschen (Chanterelles) grew under deep-shaded spruces on dark brown needles, and I might see kinglets hopping among the branches above. Chocolate brown Steinpilze (Boletus) grew under huge beech trees along slopes that led down to the heather bog. The Birkenpilze (Leccium) with their scabrous stalks grew near birches when the first leaves began to fall and the ferns turned yellow. There were never too many mushrooms in any one place, so I had a chance to move around and to be by myself for many hours. When there is a space, the mind fills it. I filled mine with the sights, sounds, and smells of the woods.

Fishing for brown trout was the most fun of all, even though I had to spend most of the time on my belly to do it. I would lie on the edge of the stream and reach over and underneath the undercut banks. When my fingertips touched the smooth skin of a fish, I reached forward bit by bit until I had felt enough and could visualize in my mind the shape and size of the trout. Then I'd grab quickly, as hard as I could, right behind its gills.

I had to hide each fish so that neither the gamekeeper nor the British occupation soldiers who sometimes also fished there (with hook and line) would know I was poaching "their" fish. I liked the soldiers. They sometimes gave me chocolate bars when they were out hunting. Also I found many of their cigarette butts from which I carefully took the tobacco for the grown-ups; cigarettes were scarce and available only on the black market.

I had no playmates and never owned a toy. Yet I didn't feel deprived. Who needs toys, after having seen caterpillars from up close and knowing they can turn into moths? In the spring I watched the brilliant red peacock butterfly with its yellow and blue eyespots on the wings, as it sunned itself after coming out of hibernation. Its spiny black caterpillars would later feed on the nettles growing along the brook. We, too, fed on the nettles, which make a good "spinach" when boiled.

I watched the warblers singing and fluttering in the shady forest as the green, almost transparent beech leaves were opening in the spring after a warm rain. I found their domed nests on the ground among the dead leaves, and I saw the tiny speckled eggs. In the gnarled willow trees by the brook I saw a tiny long-tailed titmouse disappear into a fork of a tree. The bird had entered a nest covered with lichens, and the nest looked exactly like the lichen-covered tree itself. I climbed up and examined the lichens held together by spiderwebs collected by the birds. Soft feathers protruded from the entrance hole, and with my fingers I felt tiny eggs within the nest. I especially liked the bright blue titmouse. It was a compact little bird that was very inquisitive and like an acrobat, as easily hanging from branches as perching on them. It raised its young in old woodpecker nest holes.

In the fall the beechnuts and acorns ripened, and we had another harvest. We sold the acorns to the farmers in the village, who fed them to their pigs, and we sold the beechnuts to be pressed for oil and made into margarine. Picking beechnuts was worse than picking berries because they were so small, and because Mamusha did not like it when I stopped to pop one into my mouth.

The British occupation soldiers who came to the forest to hunt were

Blue titmouse

not all good stalkers and trackers. But they had a lot of ammunition. They sometimes wounded game that got away and died later. One morning, when there was frost on the leaves, I saw ravens circling near the bend in the sandy road where I walked to go to school in the village of Hamfelde. I moved into the dense spruce thicket where the ravens went down, and there I found a half-eaten boar with coarse black hair and long tusks. The guts had been pulled out and eaten by the birds, but it still had fat on its hide and a little muscle. I got help, and we hauled the remains home to our cabin, where we cut the fat off and fried it. I'd never tasted anything so delicious.

Boar were not scarce in the Hahnheide, but Marianne and I were sometimes less afraid of them than of people. I particularly remember one incident as we were coming back from the village carrying a fresh loaf of bread with thick, flaky crust. The aroma of that bread was *overwhelming, and we succumbed, nibbling off some of the flakes of crust. They were so good that we pried off a few more, then pieces of the crust itself. We felt guilty for our misdeed, and walked slower and slower. Eventually dusk arrived, and we were still only at the bend in the dirt road where there was a slight rise. And at that place, looking up, we saw a man standing, staring at us. We panicked and bolted into the underbrush. I then led the way through the woods back onto the road far ahead, almost up to the footpath through the spruces that led us to the safety of our hut.*

Early one spring, shortly after the snow was gone, we found ourselves without food: we had eaten all of our dried berries and mushrooms, and there wouldn't be new ones for a long time. Papa and I walked into the forest. He didn't seem to know where he was going, and he didn't talk at all. We came to a stand of beech trees, but there were no nuts under the dead leaves on the ground. The new leaves weren't yet out, and the sun still shone through the branches. We walked around all morning and found nothing. When we got tired, we sat down in the sun at the base of a huge tree. It was quiet, with only a dog yapping in the distance. Suddenly Papa sprang up, as though stung by a bee, and ran off in the direction of the sound of the dog. When he came back he was carrying over his shoulders a deer that was bloody with a torn throat. He had heard and recognized the cry that a deer makes when it dies. We picked our way carefully through the thickets back to the cabin, away from the dirt roads where the soldiers might see us and take the deer away.

Later Marianne and I found the third, and best, large game animal we were to have in our five years in the Hahnheide. One of our daily chores was to collect branches for the fire. Occasionally we heard a boar crashing through the thickets, and in the fall we heard elk bulls bellowing at night. One day we saw what looked like a huge patch of brown fur in a thicket near the brook. We came closer, and it didn't move: a dead elk. We ran back to the cabin and told Mamusha, Papa, and Ulla, who did not believe us at first. But they did investigate and could hardly believe our good fortune. We hid the elk by covering it with branches and waited for nightfall.

As soon as it was dark we dragged the elk into a spruce thicket, where we could work by candlelight and not be seen from the nearby road. Through most of the night we were busy skinning it, cutting it up, and carrying the pieces to the cabin. Of course we had no refrigerator. But we could not afford to let such a treasure spoil, nor could we afford to be seen with it. Our solution was to hang some of it in the chimney, where it was preserved by the smoke and hidden as well.

We were not permitted to own firearms, but slingshots were allowed. I made one using a forked branch and rubber from a discarded inner tube. I recall the intense pleasure of stalking birds in the woods, and occasionally killing one. Having a bird in my hand gave me a chance to see details of eyes, feet, and colors and patterns of feathers that could never be imagined. We also trapped some birds using the hairs from our horses' tails. The long, strong hairs were made into stiff lasso-like loops that were pegged onto a board. Birds hopping on the board, trying to pick up grain, were entangled in the snares and held fast. We hunted mice and shrews with pit traps dug in the forest. We fried the bodies of the birds and rodents, bones and all, in the grease we got from the wild boar the ravens had shown us, and there is, in my recollection, no greater treat than mice fried to a brown crisp, except possibly pork chops (which at that time I did not remember having eaten).

We used not only the meat of the animals we caught. We removed fleas from the mammals and sold them to Dr. Rothschild, a flea specialist in London. Each kind of mammal hosted different species of parasites. Merchandising fleas didn't make us rich, but every pfennig helped. Also, Mamusha skinned the birds and mammals and made study skins that we sold to the New York and Chicago museums of natural history. However, at this time there was no mail service from Germany to other countries. Again we had incredible luck. Papa had long known Piet Hart Nibbrig, the general manager of the Bols Liquor Company in the Netherlands. Piet and his wife Nela had been frequent guests at Borowke before the war. As Hollanders, they were allowed to drive across the German border. In order to help us now, Piet picked up our merchandise and brought it to Holland, whence it was mailed to New York.

I enjoyed the hunting and trapping. It was always interesting. Before digging a pit trap for a small mammal, we searched for tiny tunnels under leaves, matted grass, and moss. It intrigued me that, as if by magic, various kinds of small furry animals had built amazingly intricate mazes and were hidden from our view, probably aware of our presence. Whatever it was they were doing, it was aimed at survival.

To make a good pit trap for mice, one must make sure that the sides of the pit are smooth and that no piece of root hangs down to afford a foothold. Mice caught in pits we made in the Hahnheide often tried to burrow down when they could not escape upward. We put a large clump of moss in one corner in each pit to give them a false sense of safety, so that they would not try to escape. Animals would hide under the moss rather than try to jump, climb, or dig out. I always ran ahead of Papa, to be first to lift the moss to see what kind of field, wood, or deer mouse or shrew might be underneath.

While we were building and tending our pit traps to catch mice, Papa showed me some small conical depressions in loose sand. They were pit traps made by an insect larva, the ant lion. These traps are designed to catch ants, beetles, caterpillars, centipedes, spiders, and other prey. The ant lion itself lies hidden in the sand at the bottom of the pit; there it flings up sand to line the walls with loose, treacherous footing. If a victim blunders past the rim, it slips down to the bottom of the pit.

Shrew, with remnants
of its last meal

I liked to stretch out on the ground and drop red wood ants into the ant lions' pit traps, watching the ant-lion larvae hurl sand up the sides of the pit. The loose sand fell back down, perpetuating small sandslides that kept the ant slipping downward, until the ant lion could reach it and pull it under the sand at the bottom. Little did I know then that thirty years later I would again be feeding ant lions in order to learn the details of their prey-catching behavior.

Along the sandy road near the ant-lion pits, I noticed many tiger beetles (Cicindelidae). Like the ant lions, these beetles also hunt insects in the sand, but they do it by pursuit. Iridescent blue-green with white spots, they are very easy to see but extremely fast and hard to catch. Their larvae are also predators, and like ant-lion larvae they live in pits in the sand. Their pits have no funnels, though. Instead, the larvae rest in the burrow with their heads flat on the ground, and they grab anything moving that comes within reach of their sharp mandibles.

There were also insects in our pits. In particular I liked the large ground beetles. Some of them were even larger than a shrew, others were tiny. They looked like six-legged knights in black and shiny armor, polished in colorful metallic sheens. They were grooved and studded with depressions and bumps of all colors. There were many species, each of them different. Even though we had very little money, Papa bought me a book on beetles for Christmas. The book showed me many species I had never seen, and the idea of starting a beetle collection of my own was born.

At first I collected only the large showy ground beetles, of the genus Carabus. Their names soon became very familiar to me. There was the shiny green **Carabus** auratus with long smooth grooves on its back and reddish legs. There was **Carabus** violaceus, who was black with violet edges. **Carabus** clathratus had raised ridges on its back, with gold and copper-colored depressions on a blackish-green background. I was particularly excited by the rare **Carabus gigas,** a giant black-studded nocturnal snail hunter that was almost two inches long. Soon I could no longer find new carabids, and I started to collect other beetles. An exciting world was opened up, and I became intoxicated with discovery. There were beautiful long-horned beetles who mimicked wasps, giant brown stag beetles, pretty red, white, and black ladybugs, bright blue snout beetles who curled up fresh leaves into wads that fed their larvae later on. I loved even the June beetles in their soft browns, handsomely marked with white patterns on the sides.

It was fun to try to figure out the subtle variations between the different species and to find the features that united others who differed widely in size, color, and shape. Also the beetles were so secretive that hunting and finding them fully occupied my waking hours. The happy idea of always being able to find new ones, given enough effort, filled me with energy and desire to explore the forests and fields in search for even more treasures.

I still keep the memory of the intense pleasure derived from finding this great variety of insects, knowing it will always be there, a prospect of never-ending delight. Aside from the memories, I have a picture of a stag beetle I drew on a small card for Papa when I was nine years old. On the back of the card is written that I had collected 447 beetles of 135 species and that I wished him a happy birthday.

Carabids, who remained my favorites, are nocturnal predators on other insects. Some catch snails, trailing them like bloodhounds along the snail's slimy tracks. Many are flightless but can run rapidly on their long thin legs. One species, the searcher (**Calosoma sycophantus**), in particular likes to eat caterpillars. The searcher has been introduced from Europe to the United States to help control gypsy moths. It hasn't done a good job, but the sight of this striking beetle with iridescent green and reddish wing covers is quite impressive.

Sometimes I daydreamed: If such fantastic six- and four-legged creatures could fall into our pits, maybe some fantastic two-legged ones, the gnomes that George (the British prisoner of war at Borowke) told me about, were also wandering about at night in the woods. But of course they would be too smart to fall into a pit. In my spare time, spent mostly in the forest, I made little moss huts for the gnomes. After all, if a giant **Carabus** is a reality, then anything is possible.

During my food-gathering excursions in the spring I became acquainted with wood pigeons that nested in pine trees. Squabs are delicious food, especially just before leaving the nest when they are still plump and tender. I hunted for pigeon and crow nests, and tied strings on the legs of the young birds to fasten them to the nest. In that way I could be sure of getting the young birds at the last possible moment before they left the nest. To hunt for nests became the great passion in my life. I loved the challenge of climbing trees to inaccessible nests, to collect the beautifully colored eggs of the different species for my growing collection. I kept one egg from each species as a trophy (carrying it down from a tree by holding it on my tongue to free my hands), and before I was nine years old I had a collection of eggs from 35 species.

I particularly liked looking for the nests of marsh birds, who seemed to be the most secretive of all. In part this interest stemmed from Papa's telling me of his expedition to the swamps of the Danube delta. There were marshy meadows along a stream a few kilometers from our hut, and in the beds of reeds mysterious marsh birds called from their hiding places. To preserve eggs for my collection, I used a pin to punch a hole in each end of an egg and then blew or sucked out the contents. I used only eggs that would not float in water (rotten and partially incubated eggs float). Papa said that absolutely no food must be wasted. I had to eat even the yolk of the tiniest egg of a sparrow. I still hunt bird nests, but I now take them home with me on film.

Some may not like this notion of robbing bird nests and sucking eggs. But I knew even then that, while the bird is still laying its clutch, if you take one egg out of a nest, the bird simply lays another. If the whole nest is destroyed, most birds will renest within days. Much can be learned from collecting eggs of songbirds and no great harm is done, unless you confuse the nest of a robin with that of an eagle, who might not renest as quickly. Unfortunately, to most administrators who make the wildlife laws, a bird is a bird. They see in black and white. The world itself, however, is colored in various shades of gray, and all the other colors besides.

Although we were always concerned with food, it was never my responsibility to provide it — not, that is, until I adopted Jacob. He was all

Jacob

mouth on a scrawny neck, attached to a pink, pinfeathery pot-bellied digestive machine with weak spindly legs that could not hold him up. Jacob was a baby crow I had taken out of a nest high up in a spruce tree. I soon loved him more than anything else, and he needed me. He had to be fed every half hour or so, and he kept me constantly on the run for worms, caterpillars, frogs, grasshoppers, and pupae of red ants. After two weeks on this diet he flushed out in a coat of black feathers, and soon he was flying around and following me everywhere. He came when called, landing on my arm or shoulder, cawing loudly in my ear to beg. Later in the year he begged for food less incessantly. He quieted down and made affectionate noises suggesting that he was happy and content. We were of the same world, despite our differences.

I had a good time in the Hahnheide. I loved spending all day in the woods, and I dreaded the idea of growing up and having to work all day. For the time being, parents attended to the more serious matters, such as making money. They tried selling stuffed mice dressed in Mickey Mouse clothes. There weren't many takers for those. Fleas and lice weren't selling well either. We never did pick enough beechnuts or mushrooms to get back more calories than we put in. But the three military horses that Ulla had caught during our westward flight were the key to a small cabin industry: bootlegging. We traded the horses to a farmer at the nearby village for several large installments of sugar beets

Peacock butterfly basking
in spring sunshine

and some milk cans. Then we hid the beets in a large pit that was camouflaged with leaves and pine needles. Papa collaborated with another refugee, a chemist, who somehow got copper tubing and made a still. We boiled the beets to extract the sugar, and the broth was allowed to ferment in the milk cans to produce alcohol. Several times a week the alcohol was distilled off. Marianne and I did our part by bringing in wood from the forest to boil the mash.

A major complication in our moonshining business was getting the schnapps to market. The market was in the big city, Hamburg. The main train station was continuously guarded by police who checked to see that none of the cityfolk were returning with food illegally bought in the country. Liquor would also have been confiscated, though most likely not for democratic redistribution to the thirsty populace.

Papa always brought his bottles into Hamburg hidden in the bottom of a knapsack. They were covered with clothes and, for good measure, topped off with several cigar boxes within which were neatly pinned row upon row of dried wasps with their legs and wings meticulously spread.

*(They were a few specimens from his new collection of parasitic wasps.) He usually got off two train stations before Hamburg, where there were fewer police guards. But one time two policemen were stationed even there. He must have looked suspicious because they followed him. As he tried to make his getaway by jumping onto a streetcar, they caught him. "Was haben sie da?" (What do you have in there?). "Sie werden lachen—Fliegen!" (You'll laugh—flies!). They lifted out some cigar boxes, and they **did** laugh. They didn't look into more than one. Having seen one fly, they figured they had seen them all.*

One day while I was by the brook catching trout, I heard a loud humming of insects above me in the gnarled spreading branches of an old willow tree. It was a beautiful warm spring day, and the sky was bright blue. The tree was covered with yellow pussy willows. Wooly black and rust-colored bumblebees were buzzing here and there. Willow warblers and pied flycatchers were hawking flies. The combination of sights, smells, and sounds gave me a delicious, light-headed feeling. Many years later during my Ph.D. oral exam at UCLA, I was asked why I wanted to study biology. I answered that it was because of what I saw and felt that spring morning in the Hahnheide. Of course this was an inadequate answer, but I didn't have a better one. How could I explain the Hahnheide, and all that led to it, to a group of five no-nonsense professors? Some things cannot be explained in three sentences.

Last year I applied for a grant to study insect thermoregulation. An anonymous reviewer of this grant aimed to criticize my proposal by saying, "Heinrich just wants to play." He couldn't have made a truer statement. Although my methods of observing nature have become more rigorous with my formal education, I do my research—my observations of nature—because it makes me happy.

Maine

At the Hamfelde village school we used to sing a rhyme that translates something like this: "We are going to America, and who comes too? The cat with the long tail, yes, she comes too." Going to America could be anyone's vague dream, but you couldn't take along your cat too. You could only leave if you were willing to sever all connections. In our five years in the Hahnheide forest we had collected no cats with long tails, and Jacob had been killed by a goshawk. We were free, we could go.

We steamed into the harbor of New York City on an April morning in 1951. The skyline ahead was a mass of jagged crags. Ten years old, I had little sense of scale or perspective. From out in the harbor I saw what looked like long lines of beetles streaming along the shore. I shivered. Would we see cowboys, Indians, rattlesnakes, and hummingbirds when we stepped ashore? What would we catch to eat?

We walked down a long steep gangplank onto a sidewalk. There were no shy and elusive hummingbirds, just flocks of pigeons and English sparrows. They seemed absurdly tame, but I didn't see anybody trying to catch them. There were no cowboys either. A man wearing a suit, and smiling a lot, was talking to Papa. I couldn't understand a word. He handed Papa, who was also smiling, a small piece of paper. I didn't know it then, but it was a check for $3,000 from the New York Museum of Natural History for Papa's catch of birds in Celebes before the war and for mice and birds we collected in the Hahnheide. It was a new beginning, but like all beginnings it started from a past.

We stayed for the first few days as guests of people called "Ernst and Gretel," across a huge bridge, out in New Jersey. What I remember most about the visit was a bird nest that hung on a wall of their house. It was a baglike contraption woven out of gray fibers. It was beautiful, and I'd never seen anything like it. I was told that it was built by an orange and black bird called an "oriole," and it came from a tree on their own street, an "elm." Gretel then showed me the birds that came to the feeder by the kitchen window. I saw familiar types — jays, finches, titmice. But these were all different from those in the Hahnheide. The jays were blue, the finches were reddish-purple, and none of the titmice were blue or had long tails. The New World, I knew then, was going to be exciting — I couldn't wait to see what other birds there were. (About twenty years later, after I had studied biology at the university, I learned with a shock that "Ernst" had been none other than Ernst Mayr, one of the most eminent biologists of our time.)

Two days later we were driven in a car over muddy roads to a farm in a remote section of Maine. I was not used to riding in a car and felt sick all the way, not noticing much, and sinking lower and lower into the seat. We stopped at a house surrounded by large maple trees. A red barn several times the size of the house stood next to it. The barn must have had a hundred or so empty cow stalls on one floor, with a mountain of decaying, weed-covered manure piled outside. The barn loft was half filled with beans in dry pods still attached to vines. There were beans as far as the eye could see. Mr. Cunliffe, who owned and would let us use this place, had another home in Florida. In return for letting us stay here he wanted us to do agriculture for him. We were to be tenant farmers, much as we had had tenant farmers in Borowke. We were instructed, first of all, to stomp on the bean pods to extract the beans. That should keep us busy for a few months. Meanwhile, Papa was also supposed to start growing a crop of potatoes.

I stomped beans for a day or so in the dusty confines of the loft, but soon I became distracted by the swallows twittering above, as they flew in through the broken window with their beaks full of white chicken feathers that they used to line their mud nests set on beams along the ceiling. I tried to figure out ways to get at a nest and look inside. There was also a pair of beautiful multicolored falcons with blue wings who nested in between the walls by one of the decayed eaves. I pried away a board and saw four nearly spherical eggs with thick speckles of chocolate brown. Cheerfully trilling little brown birds ran like mice among the rank grass and dead burdock stems that were poking through the rusting farm machinery near the barn. Azure blue birds, perched on fenceposts, called in a strange soft voice, and after fresh green grass had pushed up through the brown cover of the old flashy black and white finchlike birds sang their sparkling, cascading songs while circling with rapidly fluttering wings over dusky females hidden in the grass.

Being cooped up in a dusty barn, while the sun shone and the birds sang and the insects flew outside, was clearly not "natural" for me, or at least I would have so rationalized had I known about rationalizations. I did not fight my natural tendencies. Rather than working I was soon playing, exploring the nearby fields and forests.

Wild cherries blossomed in white profusion along the ancient stone walls surrounding the fields. Skunks lived there as well as tiny black and yellow striped squirrels, called chipmunks, who scolded and then fled to the dark recesses among the rocks. In the hemlock trees of the forest I often heard a strange and excited chatter, and I strained my eyes to find the source. For a long time the animal seemed to be invisible, but then I discovered that it was a small red squirrel who usually sat upright on a branch close to the tree's stem. Later, after finding these squirrels a few times, they became much easier to spot.

Many small, colorful wild bees of various species crawled over the cherry flowers. Occasionally I saw an animal hovering, moving quickly from flower to flower. It looked like the tiniest hummingbird imaginable, and I longed to hold one in my hand. Eventually I caught one in my insect net, where it fluttered so fast it hummed, and when I finally had it in my hand it turned out to be a moth. It was a hummingbird moth (also called a sphinx moth because of the stiff sphinxlike pose of the caterpillar). Most

Red squirrel

other sphinx moths fly at night, and I later caught them in the evening on the flowers around the house.

The neighbors were friendly. Old Mr. Cunningham came almost daily, limping across the hay field from his ten-cow dairy farm with a bucket of milk as a present. One day a man came by and asked if we were the new immigrants. Then he got out his wallet and handed Papa twenty dollars. In particular, the Adams family down the road initiated us into life as Mainers. I spent a lot of time with the Adams boys, Jimmy and Billy, hunting raccoons, tapping maple trees, and fishing at Pease Pond.

The pond was shallow, about a mile long and half a mile wide, and bordered by pickerel weed and lily pads. Floyd Adams had a rowboat hidden in the alders, and on summer evenings we went out to catch white perch and other fish. The glassy surface was broken here and there by insects dipping into the water or perch coming up to eat them. Small bats wheeled against the moon and dipped low over the water. It was always still and peaceful, with the barred owl occasionally calling from the swamp near the outlet.

In my mind the bog along the stream outlet of the pond became a precious alien world — large and hard to penetrate — and I wanted to explore it. The outlet itself was narrow and choked with lily pads and beaver dams. One could navigate only a little way with a boat, but just far enough to see turtles, muskrats, and many kinds of different birds. From deep in the interior I could hear the babel of many frogs, and occasionally a loud "ka-thunk, ka-thunk," like somebody driving posts into the ground with a hammer. At that time I had no idea what kind of animal it was. It could as well have been a dinosaur. Later I learned it was a kind of heron, the bittern.

The bog was not one homogenous mass. It had many environments within it. In some areas were tussocks of grass, with deep black mud in between. There were also dense stands of cattails and, right at the water's edge, a huge mat of interwoven vegetation where the peat moss, pitcher plants, and cranberries grew. On sunny summer days I swam alongside this mat, peering underneath into the dark, not knowing what kind of strange creatures might be lurking there. Crawling onto the mat I walked as on a springy mattress, until I came to the stunted black spruces where yet other birds lived. Many years later this became one of my study areas for bumblebees.

In the winter everything froze solid, and I could explore the area at will. I followed fox tracks around the tussocks of grass and down the outlet. I saw the red fruit of the swamp rose sticking up against the snow among bent brown blades of grass. Flocks of siskins and redpolls foraged for the seeds of the birches and alders, leaving seed capsules sprinkled like pepper all over the snow wherever they had been. I wanted to hold some of these birds in my hand, to see and feel the colors and textures of their feathers. I became like a young cat stalking mice. However, not having a cat's spring or its claws, I had to compensate and, at age ten, I was at my apogee of skill in slingshot use and maintenance. Billy and Jimmy Adams were impressed, especially after I bagged a hummingbird

Black-throated blue warbler
at nest in understory in
mature deciduous woods

at the high-bush blueberries. I wasn't supposed to shoot birds for the fun
of it, and I got a good hiding. But at the time I didn't understand what
was wrong with what I'd done. I thought that shooting birds and stuffing
them full of cotton was "Ornithology." My father had done it, and the
American museum had paid him $3,000.

   A neighbor, Mary Gilmore, gave me a big, beautifully illustrated book,
The Birds of North America. The book had numerous color plates by
Louis Agassiz Fuertes, who made the birds almost seem to flutter on the
pages. Reading about the habitats and nests of the different birds, I

yearned to see them and their nests in the wild. Each bird was a possible and separate adventure. Every time I discovered a new nest I felt that I'd unlocked some secret of nature. Finding one nest of a species usually did, in fact, provide a rough key to most nests of that species. I was beginning to "think" like each of the different birds, able to anticipate what they did and where and when the different species would most likely be found.

The best time to go looking for crows' nests, I learned, was during or right after a long heavy rain. Crows nest in many kinds of trees. But in Maine they seemed to be partial to the state tree, the white pine. They nested in the dense topmost branches. When the pine needles were thoroughly soaked and stuck together, it was easier to see through the branches. Wherever the smallest pinpoint of sky could be seen, there was no nest. There were innumerable dark spots that might be nests, and I was always looking against the sky, trying to disprove to myself that these dark spots were nests by seeing a few rays of light through them. Real nests were opaque. It was like science: first you look for something, and then when you think you have it you do your best to prove yourself wrong.

A raven, one of
many tame corvids

Occasionally I heard the drone of airplanes and saw the machines silhouetted against the blue sky. Often, on clear spring days, I heard an ominous "thump — thump – thump – thump thump thump thump-thump-rrr" from the direction of Picked Hill. Was it a hundred miles away, ten, one? Artillery practice? (The Korean War was in progress.) Eventually I learned that it was only a male partridge drumming to attract a mate. Here was, indeed, a safe and friendly place. It seemed strange that I felt a chill on seeing planes. I also felt a strong aversion to forsythia flowers, which I had once so much wanted to pick near the bomb shelter. However, I had no concrete memories of unpleasant things.

We had been used to living without a real home for a long time, but we were finally to have one again. An ancient run-down farm, the "old Dennison place," was for sale within a mile of the Adamses', on the other side of the pond. It reminded my parents of Borowke. The house didn't have chestnut trees around it, but the place was surrounded by thick maple trees and set within 120 acres of unkempt fields and woods. It also had a three-seat outhouse behind the woodshed next to the garden. I peered down into one of the holes and found myself gazing into a phoebe nest built onto a weathered beam. The nest was made of green moss and contained five translucent white eggs. The outhouse had much to recommend it (except during blizzards), and it told of an earlier time on the farm when people either liked to use it in company, or had such large families that they were forced to. Having so soon run out of patience on shelling beans, I could see why having many children might be advantageous. Nobody foresaw the consequences of overpopulation, when there was still so much space.

Aside from the wonderfully preserved and functional outhouse, the place was somewhat dilapidated. But it was cheap, and so we bought it despite its leaky roof, broken windows, and crumbling plaster walls and ceilings. Papa handed over the $3,000 and it was ours. The Adamses, Potters, Ellriches, Allens, Curriers, Cunninghams, and many other neighbors from ten miles around had a big housewarming party for us. They came in cars, station wagons, and trucks, bringing us dishes, pots and pans, silverware, saws, axes, crowbars, chairs, tables, a stove, preserves, blankets, mattresses — in short, anything and everything one might need to get a new start. Mamusha had, in the meantime, learned from the neighbors to make "home brew." It was, this time, strictly for our own consumption, and for the most wonderful people we'd ever met. After that, the neighbors returned often to chat and to drink our brew.

It took years of picking apples at the various farmers' orchards, cutting pulpwood in the woods, and working at the local woodworking shops in town before we could eventually afford to put in plumbing to replace the iron hand pump by the sink, and electricity to replace our kerosene lamps. But in less than a month I had built and installed bird houses in most of the maple trees surrounding the house. They were used by tree swallows, starlings, bluebirds, and a sparrow hawk. I again had a pet crow.

The old farm had several old hay fields in various stages of regrowth, which provided homes for woodcock, rabbit, and even partridge. The surrounding hardwood forests harbored squirrels, raccoons, and deer. These animals, along with porcupines, also fed in the ancient apple orchard that was rapidly being claimed by birch, red maple, and pine.

Under the tutelage of two neighbors, Phil Potter and Floyd Adams, I "lined" wild bees (following them to their wild hives) when the goldenrod in the fields was in bloom in August, and we all shared the excitement of finding and raiding bee trees. Many bumblebees of various sizes and bright colors foraged on the goldenrod along with the honeybees; but Floyd told me they had little honey, and were very difficult to line to their nests, so we paid little attention to them.

Phil and Floyd were curious about bees beyond their usefulness in supplying honey, but their interest was probably awakened by having had the chance to use them for that practical purpose. My interest in birds may have arisen similarly, when I searched for wood pigeon nests in order to get squabs to eat. Learning is always selective. Even some insects learn very quickly, but they only learn those things relevant to their survival which their neural circuitry has been wired to receive. We are not fundamentally different. Our neural circuitry just has many more

Frank — on a
social call

loose ends that are free to make various connections, as it must if we are
to survive under a variety of circumstances, many of which cannot be
predicted in advance.

A fox on his rounds over the fields and through the woods cannot
predict in advance if it might be best to hunt for field mice, deer mice,
rabbits, chipmunks, snakes, beetles, or grasshoppers. He must be open
to all possibilities and be able to respond to each appropriately. He gains
a broad range of skills and general knowledge by playing and by
exploring his environment. Foxes who play the most, learn the most —
and they survive. Almost anything a fox learns will come in handy later in
his life.

We too are predators, though we live much longer than any fox and
our environments are much more diverse. Small wonder play is such an
important aspect of our lives, especially when the environment is unfamiliar. Growing up in Maine I played a lot, but I never connected it to

"education." Education, in our modern conception, is usually associated with a training toward some practical end. My education was largely outside that mold. My favorite form of play in the summer was looking for bird nests or caterpillars. In the winter I loved to strike out into the woods on snowshoes, with a small backpack holding some food. I explored the swamps and ridges, looking for fresh animal tracks. I followed the tracks of a weasel from morning till late afternoon, hoping only to get a glimpse of the pure white ermine with black-tipped tail or to see the fur, blood, and feathers where it made a kill. To see these things was a great reward. While on the trail I often built a fire in the snow, to roast some meat on a stick, and to dream about the stories I had read by Ernest Thompson Seton and Jack London.

If evolution has found it important enough to have complex animals do things totally irrelevant to their immediate survival, then perhaps there is ultimate survival value in having some of us collect flies and beetles, if for no other reason than the fun of doing it. My father, a specialist in ichneumon wasps, lost no time in pursuing his hobby of building up a collection of the American ichneumonids. It was, at first, only for his personal pleasure. There was no market for these wasps, and Papa did not get paid. I took no small part in helping him collect the insects, and it occupied us year round. In the winter we looked for hibernating wasps (all females) by chopping open decayed stumps and logs in the woods, and by looking under moss on rocks and trees. In the spring, summer, and fall we hunted with a hand net, searching in all conceivable habitats. Even after years of hunting, we still kept finding "crackers."

A cracker was a wasp represented by less than six specimens in Papa's collection, and finding it was usually rewarded with a biscuit or a piece of candy. The reward system made the task something like an Easter egg hunt. Some of the wasps themselves were gaudy, brightly colored in various shades of red, brown, yellow, blue, and black. Many had white markings on antennae, legs, and thorax; many of the hundreds of species had only slightly different markings; yet males and females differed greatly in shape and markings. The variety was incredible.

Weasel in snowstorm

These beautiful and important creatures, parasites that helped to keep the populations of moths, butterflies, flies, and beetles in check, were all around us. Yet they are so secretive and unobtrusive that they are known to only a handful of people. The family Ichneumonidae is one of the largest of all animal groups. It includes more species than mammals, birds, reptiles, and amphibians combined. There are more species of inchneumonids than there are in any other family of insects. To most people with some interest in entomology, an "ichneumon" means only a large delicate wasp (**Megarhyssa**), with a fantastically huge ovipositor several times its body length, that parasitizes the sawfly larvae in wood. But these particular wasps are only a tiny subgroup of many subgroups that all specialize to parasitize different hosts. Once I saw one rare species appear in the thousands during a caterpillar outbreak. The next year the caterpillars were nowhere to be found. Papa was constantly describing new species. He even named one I captured after me: **berndi.** I was immortalized by the parasite of a moth!

I did not always collect ichneumons willingly in the summer, since it competed with fishing and watching birds and looking for their nests. Also, I felt self-conscious being seen walking around with an insect net in my hand. (A sportswriter said of Bill Rogers, one of the world's best marathoners, that he is "a tough competitor, **even though** he was once a butterfly collector.") Finally, however, I dared to walk within sight of farmers, millworkers, and lumberjacks with my net, and after that I felt I could brave anything.

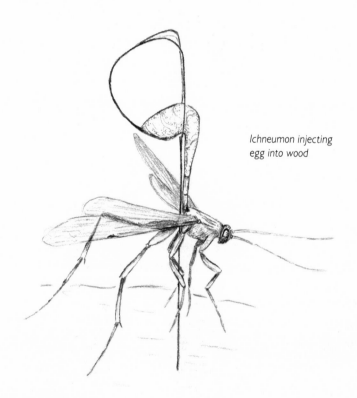

Ichneumon injecting egg into wood

Papa eventually amassed a very large collection of the ichneumons of northeastern America, and he got jobs related to wasp taxonomy with several state departments of agriculture. For personal reasons, though, he wanted to see his old collection in Borowke, which he had buried in metal trunks in the woods when the war broke out. The map to the burial ground of these trunks was in his head, and twenty years after the war he wrote the authorities in Warsaw, telling them where the collection lay buried. They dug it up, unsealed the trunks, and found the insects in excellent condition. They claimed them for the Polish National Museum. However, as agreed in the bargain for the map, they sent Papa some of the specimens to study here, on temporary loan.

I was learning more about the animals and plants in the woods and fields of Maine all the time, and I came to feel more and more a part of the new environment. It was becoming my home. Special details in the habits of birds and insects connected my mind to the place. School, however, was something to be endured from morning through afternoon, until I could again go into the woods to observe and explore.

The urge to spend time in the woods was especially great during my years at the Good Will Farm and School, in Hinckley, Maine, where I boarded while Mamusha and Papa were on collecting expeditions to Mexico and Angola. Good Will was a home for disadvantaged boys and girls, and it was a largely self-sufficient operation based on 3,000 acres of fields and forests. I helped to earn my keep by milking cows, haying, weeding, slaughtering chickens, chopping trees with an ax for the winter's fuel supply, cooking, washing dishes, working in the campus store and laundry, carrying mail to and from town, to name only some of the tasks that occupied us from early morning to dark on every day except Sunday (which was devoted to Bible study).

I was always an outsider. Every morning at convocation, and three times on Sunday, we went through rituals that seemed to have something to do with beings and places I had never seen, heard, or touched. The grown-ups talked with conviction and confidence about them. And they somehow made theoretical connections between them and abstract concepts. They could "see" what I could not see, and my mind was left groping in still air. I felt lost, afloat, mentally incompetent.

These introductions into unseen worlds often left me squirming and unsure — so I retreated to what seemed familiar. I fled into the woods. I could sit among trout lilies under a maple tree just beginning to leaf out, while a red-breasted nuthatch hammered in a rotten stub of a limb above. I knew what it was doing, hammering out a hold. Every few minutes the mate called nasally from a nearby balsam fir, and the bird stopped its work, poking its head out of the hole to drop a beakful of white shavings that drifted down on me. Then the birds exchanged places, and the hammering resumed. In a week the hammering would stop, and the hole contained a snug dry nest of green moss lined with feathers. The deep nest cup could contain five tiny brown-speckled eggs, which would eventually become other nuthatches. These, to me, were logical connections. I might not be able to understand them all, much less ask intelligent questions, but I could still get close to these wonderful things by watching them, listening, and thinking in terms of what I had found out for myself.

I especially treasured the times that Phil Potter, our neighbor from home and my old bee-hunting companion, came to take me fishing. We camped one spring near a brook, by an old abandoned apple orchard on some hills. We would fish for trout early in the morning. Meanwhile, as we were finishing our baked bean supper by the embers of the campfire, I heard above us a loud melodious twittering unlike anything I had ever heard before. I looked up and made out a mothlike silhouette with rapidly fluttering wings against the darkening evening sky. Phil said it was a woodcock. He had once, on another fishing trip, almost stepped on a hen as she sat incubating her four pinkish olive-spotted eggs in the middle of an unused lumbering road. This was a cock above us, in the sky, doing his courtship dance.

Most of the ground was still covered with melting granular snow, and the cock landed in a patch of bare ground on the hillside among the apple trees. By then I was already hiding under the branches of one of the young pines that would eventually claim this land. The robin-sized bird with his absurdly long beak was walking back and forth in front of me in the matted grass. Every two seconds or so he emitted a soft gurgling

Woodcock returning from a display flight

hickup, followed immediately by a loud quacking sound. After a few minutes of hickups and quacks he suddenly launched himself into the air, like a giant hummingbird with whirring wings. He flew straight out across the orchard and began to circle, encompassing not only the orchard but several acres of forest as well. Higher and higher he went, twittering loudly (his wings have specialized feathers to make the twittering audible). The circles got ever-smaller, until the bird was about 300 feet up, a mere speck against the moon. The twittering stopped, and then he burst forth in his real song—a series of liquid bell-like chirps that cascaded over the forest and the hills. Faster and faster, and louder, came the chirps in a series of a half dozen or so per second, rising and falling in inflection. He kept on descending, zig-zagging as he dove in a series of arcs, all the while twittering still faster and louder. And then after several seconds of silence, he fluttered into view again and landed close to where he had launched himself. He strutted, elevating his short stubby tail while gurgling softly and quacking loudly. If this display, which can be heard for close to a mile, did not impress a female, then I don't know what would. It certainly impressed me, and I can't remember a spring in Maine after this when I have not lain under a young pine tree in an abandoned field, taking in the woodcock's show. No wonder it was so difficult for me to concentrate on studies.

I didn't think I was ever suited to succeed academically. All I knew was that I wanted to make my living close to nature. When it came time for college, I enrolled at the University of Maine majoring in forestry, having vague notions of walking in the woods and learning about the interactions of plants and animals in their natural environment. In the summers I worked at a woods camp with French-Canadian lumberjacks in the Allagash wilderness region of northern Maine. Most of the lumberjacks went home to Quebec on weekends. To pay my way at the university, I stayed in camp to run a trapline catching shrews to be sold to a biological supply house for study skeletons. I didn't use dirt pits, as in the Hahnheide, but sunk narrow-necked glass jars baited with cheese or meat into the soil.

After I returned to college one fall, I realized I was not being taught what I wanted to know, but only trained to fill a niche in the job market. Facing courses in firefighting, surveying, and log scaling, I came to the conclusion that I was on the wrong track. Luckily, my parents were about to leave on their second expedition to Africa to collect bird skins for Yale. They offered me a chance to come along and work for them for one year as a birdhunter-taxidermist. Perfect timing.

Tanganyika Bird Hunt

Far ahead, in the dark undergrowth, was a flicker of movement. It vanished in the fraction of a second. My eyes strained in the deep shade. The movement was not far away. I checked to make sure that the 32-gauge auxiliary barrel was in the left side of my 16-gauge double-barrel shotgun and that it was loaded with a shell. The right side held the 410 auxiliary barrel for a slightly longer-range shot.

Sweat was rolling down my forehead and my khaki shirt stuck to my back. The forest floor seemed like the bottom of a musty pool filled with decaying vegetation. All the light, and the green, was above me. I was somewhere down near the roots. But it was not always easy to determine where the plants ended and the soil began.

We had only hunted a few weeks in the hot Pugu lowland forest near Dar es Salaam in Tanganyika (now Tanzania), and we already had 170 different species of birds. The one ahead of me sounded different. I had not heard it before.

There was a rustling to the right: three elephant shrews, a form of life found only in Africa. They are rat-sized creatures with long pointed snouts and long naked tails. The rustling stopped, and then they drummed their feet on the ground. Were they communicating something among themselves? Were they sounding an alarm?

The bird sang again. It was a soft, plaintive two-note call, the first part a short whistle with an upward inflection in pitch, followed by a downward trill that ended low and wavering. My eyes focused on a slight movement—a small bird. The gun came up against my cheek, and the finger pulled the trigger almost reflexively, and the bird came fluttering to the ground. I sprang from my hiding place and crashed ahead as fast as I could. I might have to search for him under every leaf starting from where he fell, in ever-widening circles, till he was found. But I was lucky this time. A small bird with partially outstretched wings was dead on the brown leaves on the ground. He was a beautiful gray-green above, with a pinkish breast. It was the Sharpe's akalat, **Sheppardia sharpei**, which we had not seen before.

As I leaned against a tree, I again heard the many bird voices of the forest. There was the soft cooing of doves, the weird loud bark of a Livingstone's turaco, and from the distance came the loud braying of a silvery-cheeked hornbill, as well as the raucous cackle of a crested guinea fowl, "Goaooak–ka–ka–ka–ka–koaaak." There was the twitter of small emerald sunbirds among the flowers of the tree above me, and somewhere in the dense foliage of another forest tree the green tinkerbird (a barbet) repeated its "tik–tik–tik." The forest was a babel of many voices. I listened to them, trying to pick one out that was still unfamiliar.

The distinctive whistle of the black and orange dark-backed weaverbird (**Symplictes bicolor**) sounded from the ravine below. At this time these weavers were traveling in flocks with many other birds, and the whistle suggested an opportunity to see new birds. I heard the repetitive and shrill warning calls of the weavers. Maybe these birds had found an owl. There might be a bird I had not yet seen among them, and I began another stalk. The vegetation did not allow me to pass through easily, and in some places I had to crawl on my hands and knees, inhaling the musty smell of decaying vegetation and staying alert for puff adders that lie coiled blending in with the leaves.

The birds continued to scold, but there seemed to be no owl, their usual target. When I saw what was causing the commotion, a chill ran down my spine. A large specimen of the deadly green mamba, known to be very aggressive, was slithering above me through the branches. I saw few snakes elsewhere. This one could easily have been mistaken for a vine. Several birds were hopping near it. The vegetation was dense, and I had only brief glimpses of birds. There was a flash of white — the breast of a Lanius bush shrike. I let it pass. A flock of greenbuls was slinking through the leaves on the branches. They were the mountain greenbuls, but one of them seemed to hop closer to the trunks of the trees than the others. There was a moment's silence following my shot. But the babel resumed soon after the greenbul tumbled. It looked slightly different — maybe another species?

I heard a turaco, and monkeys were scolding somewhere. It was mid-morning and getting warmer. Perhaps the birds would now be going to the cooler places, down into the washes. I followed, using a narrow game trail that led to a small clearing where there was a colony of yellow weaverbirds. The birds were flapping their wings while dangling from their nests, which looked like small brown grass bags hung by slender stalks from the swaying twigs above. A snake would have difficulty reaching them.

The game trail continued down the ravine. There was a soft crackling noise like that of fire going through grass, but much softer, to my left. Soon hordes of beetles, centipedes, spiders, caterpillars, and cockroaches came scurrying over the path from the direction of the rustling noise. The "siafu" (Swahili for army ants) were on a raid. Secretive ground-dwelling birds should come to feed on the fugitives attempting to escape the ants. In a short while I saw a **Cossypha** flutter nearby, a beautiful thrushlike bird with a white stripe over its eye and a yellowish breast. I recognized it from the book I started studying after we boarded the **African Moon** in New York harbor, months back. The bird flew away, but others would come in time.

The ants appeared in a broad front. Like a writhing blanket they covered everything. The blind millions climbed the trees, spread out onto the limbs, and then dropped from the leaves. The front of the mass moved forward partly from the pressure of those behind. Here and there phalanxes of ants encircled groups of helpless victims, reminding me of "the Kettle" we had been caught in during the war. The surrounded caterpillars and cockroaches were dismembered to become a part of the colony's metabolism. Only the winged insects escaped.

Numerous ground-living birds were attracted to the insects scurrying away from the advancing front of army ants. But still more came to termite mounds at that one hour or so in the year when they release their winged reproductive caste, the alates. On one of my bird hunts I came upon a termite mound where the winged alates were rising like a plume into the sky. Swifts that I had never seen before made passes like dive bombers over the mound. After one pass they wheeled around, twittered loudly, and came by for another run. I was able to shoot species that nested high up in tree holes and normally are almost impossible to get. At one mound I saw two large brown steppe eagles that perched directly on the mound and picked off the termites one by one as they came out before taking wing. At a nearby mound the alates fared little better. I first noticed the mound because of a black drongo who landed on it. In less than half an hour, weaverbirds were in the nearby trees, making occasional forays to the mound.

Cisticulas came from the grass. A blue roller was attracted. Swallows came. The crowd of birds seemed to be eating most of the insects that came out of the mound. But more and more alates came out, while the numbers of birds stabilized. Within a few more minutes a cloud hid the sun; still the termites kept coming, only now they were not flying off after emerging. Instead they piled up in a writhing mass on the mound, spilling by the thousands off the side onto the ground. I don't know why they didn't fly as the birds closed in. Perhaps it was too cold for them. Perhaps the termites that monitored the time and the weather, and then determined when the alates should be released, had miscalculated. There might be limits to what a termite can do, but from the little I had learned of what there is to know about them I was (and still am) hesitant to say what those limits might be.

After I got twenty birds it was time to go back to the tent to spend the rest of the afternoon skinning and stuffing them so that they could *become study material in the cases of Yale's Peabody Museum. Some-time — maybe next year, maybe a hundred years from now — someone will use these study skins and the data we collected for each one. The birds are of benefit to us in many ways. It is strange that, even though there are laws against killing birds, there are no laws against destroying the environment without which they cannot exist. One could spend a lifetime shooting birds in these hills without putting an effective dent in their population. Even if one did, the population would rebound in a year or so: the destruction would end with the act. But when you destroy habitat, you kill future generations. If a lumberman were to come in here with a chain saw, he could cut down all of the large trees in one year. Many of the birds would then disappear, and they might not come back until a hundred years later, after the trees have regrown. If the forest hills become isolated, by being surrounded by corn fields and settlements, the birds might be gone forever.*

Even though birds can and do fly long distances, they each inhabit a very specific "home," as I had learned in the Hahnheide and in Maine. One bul-bul lives only among the lianas close to the stems of the large trees; another lives only in the tree crowns; still another stays close to the ground. Each "knows" its habitat, its home. It has appropriate responses to escape the predators there and to find its food. The information to make these responses is coded in its DNA, and superimposed upon that is more information encoded in the nervous system. Home is where one's responses are appropriate. I am not yet at home in this environment. I am an outsider, I analyze. I still don't know which of the many stimuli I receive are relevant to my survival. I have to be open to them all — which is the price of being away from home.

In the evening Mamusha fried some of the skinned birds for supper and served them with rice. I then crawled into my puptent, onto the moldy mattress, listening to the insects and the night birds for awhile, to be awakened by the chirring calls of the doves at dawn.

We moved on to the Uluguru, Usambara, Pare, and Ukuguru mountains. Up in the mountains it was cool and it rained often. I could see almost no patches of sky through the dense canopies of the giant trees.

The canopies were often shrouded in fog, and birds were not easy to spot. After a few months in these forests I knew most of the birds. Many of the same species existed in the different mountains, but there were often slight differences between them, which interested the zoologists. The small differences foreshadow the differentiation of species, and they might provide keys to why or how species evolve.

The thornbush steppe, only a few thousand feet lower in elevation, was an entirely different world. It felt much more hospitable, perhaps because I could look up and see the sky, and watch the bateleur eagle spiral. It was pleasant to walk straight without having to push through dripping wet brush or to stoop and crawl on my belly through tangles and vines. It was easy to find birds and to follow them.

The flat-topped acacias were especially beautiful. From their branches hung weaverbird nests of many kinds, as well as hollow logs put there by natives to attract wild bees. The bees were more aggressive than any I had seen before. For millennia men and honeybadgers had raided bees' nests. Some colonies had defended themselves more aggressively than others, and these colonies were left alone; they reproduced, making more copies of themselves. We, and presumably the badgers, prized a nonaggressive bee, and as a consequence we now have a very aggressive creature. That is what lack of information and planning can do.

Among the acacias were stunted shrubs, almost all with spines. From what were they being protected? There were huge euphorbia trees superficially looking like overgrown cactuses. Why did the two totally unrelated plants look so much alike? (In America the only euphorbia I knew was a small creeping garden weed.) Perhaps like the vines with huge barrel-shaped stems, they were adapted to store water. And scattered between the acacias were giant baobab trees with big white flowers that opened at night and were visited by bats. Aloes with long red tubular flowers grew under the baobabs and attracted sunbirds of brilliant colors. Mutualisms had evolved between the animals and the plants.

The sunbirds looked very much like hummingbirds, although they did not hover. Some of the flowering acacia bushes also attracted scarabid

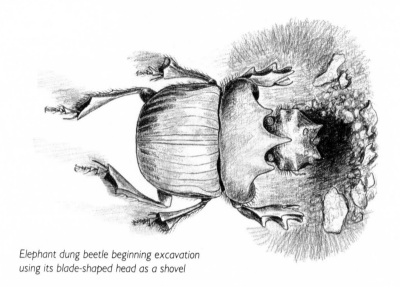

Elephant dung beetle beginning excavation
using its blade-shaped head as a shovel

beetles with shiny green elytra (wing covers) covered with white, yellow, and red markings. Despite their large size, these beetles took to wing quite often and hovered around the trees like giant bees. Meanwhile, other scarabs fed not on pollen and nectar but on the dung of mammals. They collected it for their young, rolling it in a round ball along the ground to some unknown destination. Sometimes there was a pair of beetles, with one pushing and one pulling, or one pushing and one following. Somewhere they would dig a pit to bury the dung, to eat it, to mate there, and to lay eggs and let a larva complete its development in the dungball. The dung rollers always seemed to be in a great hurry. Perhaps they had to make a quick get-away because of the guinea fowl, hornbills, bustards, and various kinds of starlings that often intercepted them at the dung. The tenebrionid beetles that look like rough pebbles, and are almost as hard, lumbered along slowly, occasionally feeding on dead plant debris. When touched they either feigned death or pulled in their legs like a turtle retreating into its shell. Toughness is a good defense where speed is lacking.

Huge clay mounds of termites stood among the acacia trees. Tiny white workers, who looked like delicate white cockroach nymphs, trav-

eled in columns from these mounds inside tunnels of clay they had constructed on the ground surface. These enclosed walkways led to fallen trees, branches, and other feeding areas. The termites seemed to try to avoid the light, but after the seasonal rains, the mounds issued thousands upon thousands of winged forms that fluttered on weak wings straight up in columns. Birds of all sorts gathered in great numbers at these mounds. There were weavers, cisticolas, rollers, starlings, barbers, swallows, swifts, and sometimes even birds of prey. The short synchronous flight of the winged termites was their elegant evolutionary solution to the problem of predation: safety in numbers.

After they had flown, the winged termites dropped on the ground. Soon one (a male) would find another (a female). With a quick shrug they shed their wings, breaking them off along predetermined lines of weakness. Most of these individuals would almost certainly soon be eaten by lizards, birds, or other insects. Only the very rare pair would survive long enough to be able to found a new colony that might grow into another termite mound.

The thorny scrub was inhabited by many animals besides birds and insects. There were dik-dik, gerenouk, and other antelopes. Lions roared at night. Their deep guttural rumbling seemed to shake the ground. There were giraffes, porcupines, and mice with spines. I presumed the acacia had spines for the same reason that euphorbias had poisonous sap, so they wouldn't be eaten. The giraffes eat the acacia anyway. Perhaps some of the other browsers were less able to share the feast.

Few could hide in the thornbush steppe. The snake could not escape the hawk and the mongoose simply by hiding, for there is no place to hide. It has to be quick as well as poisonous. Not only was most of life there in direct contact with the harshness of the elements, but it was also in contact with itself. The conflict was out in the open, and offensive and defensive strategies both seemed to be carried to great sophistication, with little tolerance for letting down the guard and experimenting. Life had to conform to a specific design. Nietzsche said that, in human society, stringent conditions produce a society that bristles with defenses as it tries to preserve a specific type. Later, when the society is well-established and secure, it tolerates variation among its ranks and elsewhere. He went on to say: "A species comes to be, a type becomes fixed and strong, through the long fight with essentially constant **unfavorable** conditions."

Ecologists have recently come to similar conclusions about evolution. But evolution is ultimately based on the sum total of individual interactions and experiences. The sequence of events leading to individual experiences can be extremely complex. Anyone's life shows that. Still, there is an underlying logic that transcends individual interactions. It is not so simple or mysterious as we may be sometimes led to believe, as in the following conversation from the 1982 MGM movie **Diner:**

> "And now they have this theory called 'evolution'." "What's that all about?" "It's about these two amoebas that lived in a swamp a long, long, time ago. The amoebas had a fish, and the fish crawled onto the land, and the next thing you know there was this guy standing on the street corner yelling 'taxi'." "It doesn't make sense to me."

Evolution is history. It is not a concept that can be shown with mathematical precision in a formula. Any attempt to do so results in absurdity. The truth of the concept of evolution rests on so vast a body of knowledge that few living people can encompass it all. Yet, without evolution as a central structure, there would be no modern biology as we know it. The mind may resist making the leap from the fish to the man standing on the street corner, but it is more prepared to do so after having appreciated the incredible diversity of life.

That diversity is almost too great to be grasped intellectually. I think it has to be felt to seem real. Academic arguments on the morphology, behavior, and ecology of butterflies in a tropical rain forest, for example, do not really convey them as a concrete reality that can be savored in other than an abstract sense. By exploring these different habitats and these different mountain ranges, we were, in effect, examining biological diversity. We wanted to find the minimum differences and minimum areas that had historically decided the life and death of various types.

Just as African wild buffaloes are not compatible with humans, except in a very large area, so certain birds were thought to be incompatible with others, especially on isolated mountain tops.

Papa wished to have birds from the higher reaches of Mt. Meru. At that time there was no road up, and I hired a guide and porters at the foot of the mountain. Our column marched through dark forest and then through tall heath thickets as we neared the top. There were many wildflowers. I was surprised to see the same kinds, but in slightly different form, that I knew back in Maine: blue irises, pink impatiens, yellow daisylike flowers.

Buffalo

When we finally walked down into the crater, however, we were in a world totally unlike any other I had seen. We came to a small knoll of short thick trees surrounded by meadows with flowers. To one side of the knoll was a small pond with muddy shores trodden into a quagmire by elephants, buffaloes, and rhino. The forest just outside the grass perimeter looked dark and foreboding. The gnarled trees were laden with thick green moss on which small vermillion orchids bloomed. The branches were adorned with light green lichen. Small pink and blue flowers dotted the verdure of the moss on the forest floor.

I felt a tense excitement that was perhaps related to experiences of the week before in the forests further down this mountain. One day as I was peering up trying to discover a bird calling from the thick foliage, I caught an unfamiliar smell. A few moments later there was an unfamiliar sound as well. It was the thumping feet of a rapidly approaching rhino, which I quickly identified and fled to find my birds elsewhere. I saw elephants routinely, but they were surprisingly hard to see in dense underbrush. One day a badly scratched and bleeding African came into our camp; an elephant had induced him to speedily ascend the closest tree, a thorn tree, where he spent a leisurely afternoon. Several days later a guide was showing me up one part of the mountain along some deeply rutted game trails, when a big buffalo bull with tattered ears charged out of a thicket and gored my companion through both thighs with a long set of polished horns. Then he crushed the man's rib cage and, possibly supposing to finish the job, he charged me next. Luckily, I was fortunate to have had several seconds' head start.

As we now examined the abundant sign of the big animals all around us in the Meru crater, I wasn't deceived by the stillness. While stalking for birds in the forest, I was ever mindful that the forest contained other stalkers as well. The constant intent listening made the silence almost deafening, and the thick opaque fog that hung heavy and low seemed to magnify the presence of these unseen animals. I saw few birds. A turaco occasionally called from deep in the forest. It was an eerie place.

The cool dampness under overcast was not propitious for insects either, particularly wasps. Yet on the ferns near the crater floor there *was a large assortment of ichneumon wasps. Ichneumons are usually alert and fast, but all of these could be easily caught by hand. Perhaps they had been flying in sunshine up in the canopies of the forest lower down, searching for the strange-looking caterpillars of sphinx and saturniid moths that I would probably never get to see. The caterpillars could mimic snakes and bird droppings and debris, or they could have poisonous spines to protect them from birds, but they had little protection against the wasp specialists. The wasps were thriving here, despite all their hosts' best defenses. Perhaps a warm wind had swept up the slopes, and the wasps came with it. Carried up to cooler air, they may have hit a dead air pocket over the crater, and then their cooled flight muscles, no longer able to generate sufficient power, became useless. And so they settled, immobilized by cold.*

There were hundreds of them, of many kinds. Each kind undoubtedly parasitized a different host, and nobody could know what parasite belonged to what host. Most of these species had not even been named. Several questions came to me: How did the wasp know to which host it belonged? And why do most of them specialize? Do they choose a kind of caterpillar that has a particularly effective defense? A caterpillar that is likely to be eaten by birds is not a good place for laying eggs. But a closely related group of wasps, the sphecids, have gotten around that by burying their parasitized caterpillar in the ground and paralyzing it with poison so that it doesn't crawl away to be eaten by birds. There is no soil available for burying things up in the forest canopy, but some wasps have gotten around that also by making small coffins of clay for their caterpillar prey.

After dark, as we huddled round our constant fire in the small knoll by the pond, we heard the sloshing of heavy mammals at the water. Our mist nets set to capture birds captured many dung beetles instead, when the nets were not torn to shreds by the large game animals. When elephants passed close by, we caught some beetles that were the size of sparrows. These large insects, unlike the wasps, were obviously not incapacitated by the cold, even at night. Near fresh elephant dung by the waterhole I saw the earth thrown up, as it is at the end of mole tunnels.

Beetles had done it. Aside from being agile flyers, the elephant dung beetles can bulldoze straight down over three feet. The female beetle goes to the bottom of the hole and makes balls from the dung that the male carries down to her from above. As with the dungball rollers, the dung serves as food for the adults as well as for the developing larvae. The beetles' action on their environment affects other organisms. With their tunneling the beetles aerate and fertilize the soil that promotes the growth of plants and places for new seeds to sprout. The plants in turn support other insects.

I knew enough science to realize that all living things are made out of the same chemicals that abound in ocean and rock, and that these molecules obey the same laws of physics and chemistry in living systems and wherever else they are found. But as I sat in this wilderness, I also knew that there had to be more. It boggled my mind that what I was seeing was how nature itself arranged every leaf, every insect, every bird, after this crater had once been a fiery inferno. It seemed like a sacred place for tunneling into the natural order of things, and even now, twenty years later, I still feel that wilderness almost as if it were with me. I made up my mind then that I wanted to understand the mechanisms of life. But, to do that, I had to start at the basic level of biological organization — biochemistry, genetics, embryology, cell biology, physiology. The sensual, overwhelming **experience** of nature would have to be put on hold.

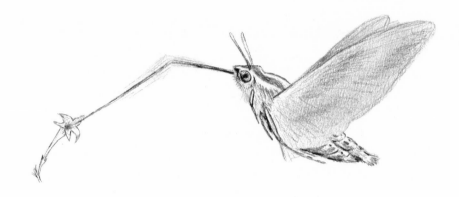

The Thesis Hunt

The protozoa called Euglena are difficult to define as plant or animal. They have features of both. Under the microscope the organisms move rapidly, propelled by whiplike tails, or flagellae. When in the dark they can absorb nutrients and metabolize them by some of the same intricate metabolic pathways found in animal cells. On the other hand, when these organisms are put into water without organic food molecules, but in the light, they miraculously turn green as they become packed full of chloroplasts like those from any respectable plant. With the chloroplasts they can synthesize their own food energy out of the carbon dioxide they take in from the atmosphere.

Professor James Cook's laboratory in the basement of Coburn Hall, the ancient zoology building at the University of Maine, was filled with Euglena housed in Erlenmeyer flasks. The flasks in the light contained a green broth, as millions of Euglena crowded into every milliliter. The Euglena cell suspensions from the dark, in temperature cabinets, were a pale transparent white. Other Euglena were held in temperature-controlled chemostats, where accurate pumps delivered nutrients at rates precisely adjusted to maintain specific population densities. There were centrifuges, an autoclave, a Beckman spectrophotometer with a recorder, a large electronic box with a fluorescent screen, called a Coulter counter (for counting population densities of the protozoa), plus the usual shelves of chemicals and glassware.

I was majoring in zoology now and helping out in Cook's lab. By the end of the summer I had graduated from washing dishes to mixing nutrient media, pipetting cultures, making cell counts, and doing enzyme assays. I was even starting to wonder about many of the differences I was seeing between cells grown on sugar and cells grown on the salt of acetic acid, and so I decided to stay on for a master's degree. Despite my inclination to be outdoors, I felt I had to work in the laboratory to start learning the roots of biology. My thesis problem was on deciphering the metabolic pathways whereby these two entirely different food molecules might be used by Euglena, and how or why the different food molecules so drastically affected the cell's respiratory rates, culture acidity, and maximum population density.

Before I could get involved in this research, I did much reading on the biology of Euglena. I saw these organisms through the minds of others, in their brilliant experiments and insights. The beauty of Euglena was inside of them, and much of that beauty could only be seen with the aid of a laboratory and a sharp mind's eye. The seeing is in the head, which is where the excitement is. I rarely looked at them through the microscope; I was more like a blind man when it came to seeing the biochemical machinery in Euglena. But after a while some parts of that internal beauty lay tangible and exposed, as if in the palm of my hand. Perhaps beauty and the structure and functioning of nature, in each of its many dimensions, are one and the same thing. Beauty is not only an inherent feature, like the molecules of Euglena, the feathers of a bird, or the trees of a forest.

Within two years I had published or helped publish four papers in leading scientific journals. My initial success encouraged me to continue on for a Ph.D., and within another year, in 1966, I was at UCLA, focusing more and more narrowly on cell physiology, working this time with Tetrahymena, another kind of protozoa.

I broke the Tetrahymena cells open to spill their contents by swirling them in a detergent that ruptured the cell walls. Now I had a soup of cell fragments, sugars, fats, amino acids, proteins, and many other chemicals. To isolate the DNA out of this soup of cell constituents and detergent, I added phenol and shook the broth until it was white and *creamy. The DNA separated out, and I gently twirled the coils out of the beaker onto a glass rod.*

I was not equipped to handle examination of the sequence of bases that made up the DNA code. For the time being, I only wanted to know if the "tape" lengths, or the overall composition of bases, varied between possibly diverse populations of DNA molecules within the same cells. One of the ways of separating different DNA molecules was by their density-dependent rates of sedimentation under a strong centrifugal force.

I took a small sample of DNA downstairs to spin it for a day in a cesium chloride gradient on the Beckman ultracentrifuge. I waited. Would I get a satellite peak or not? My densitometer reading said I didn't have one. Perhaps the machine didn't work properly. Perhaps I didn't spin long enough. Perhaps my sample wasn't pure. Perhaps I shook it too long and broke the DNA strands. Or did I do everything right and was there really no satellite DNA? The only way to find out was to do it again, and again. Same results. Surely there is supposed to be a satellite if the mitochondria have their own DNA. Like any art, biology takes skill. It is not always easy to do all of the mechanical details just right in the effort to bring any one intellectual idea to life. It is disquieting when you don't know whether a certain result is real or just the result of fumbling.

The ultracentrifuge was a rare and expensive machine. Many professors and graduate students at UCLA had to use it. A week passed before it was free for my sample. In the meantime I noticed, beyond the phenol fumes surrounding my workbench, the red-tailed hawk soaring up against the sky, over toward the canyon. I wondered where it was making its nest, but I didn't have time to look. I was studying physical chemistry and calculus, and that was taking most of my time and all of my patience. I realized then I was not in the right field. Certain keys to an understanding of "life" would indeed be found in physics and chemistry. But maybe there was more. Some of the great revolutions in biological thought were made by a naturalist who was a beetle and rock collector and by a monk who liked growing varieties of garden peas.

One of the beauties of the UCLA zoology department was its diversity. If you couldn't find a home to suit your individual needs or wants in one lab, then you might find it down the hall. There were people working on molecular biology, cellular biology, whole-organism biology, and ecology. There were also world-renowned people working on insects, fishes, birds, mammals, and vertebrate fossils. Still others studied not specific organisms as such, but the adaptations of organisms to common problems encountered in the environment, such as those relating to moisture, temperature, and pressure.

One of these was the eminent physiological ecologist George A. Bartholomew. Bart was only slightly wary when I told him I wanted to switch areas of study. He cut short our first meeting by telling me to come back in a week with six different potential Ph.D. thesis test problems. "Then we'll talk," he said.

During this time a paper by Phillip A. Adams and James R. Heath in the Journal of Experimental Biology caught my eye. The authors claimed that the white-lined sphinx moth stabilized its body temperature near 37.8°C (100°F) over a wide range of air temperatures. This seemed incredible. I had learned that the primitive birds and mammals maintain lower body temperatures than the more advanced ones and that the still more primitive reptiles and amphibians are cold-blooded. Moths and other insects are considered to be very primitive indeed compared with vertebrate animals. Thermoregulation by internally generated heat was only known for the highly advanced birds and mammals, and it was unknown even in the lower vertebrates, the amphibians and reptiles. Insects, of course, are not lower on the phylogenetic scale than humans. They are merely an entirely different line that has advanced along its own trajectory.

Sphinx moths, sometimes called hummingbird moths, are large narrow-winged insects that superficially resemble hummingbirds when they are in flight. They and hummingbirds are a remarkable example of what biologists call convergent evolution. They have both become adapted to do the same things: to suck nectar while hovering in front of flowers. They could also have converged in some aspects of their physiology, such as thermoregulation, but such a prediction does not mean that it is true. Some of the methods used in the experiments on moths were less than clear. I felt that the claim that they regulated their body temperature should be reexamined.

It was in 1964 that Adams, an entomologist from Fullerton College in California, first approached Heath, from the University of Illinois at Urbana, to study some sphinx moths he thought might be keeping themselves warm by shivering, using their flight muscles in the thorax. Heath was a well-known authority on thermoregulation in vertebrate animals, and he agreed to help investigate the problem. The species of sphinx moth that Adams showed Heath was **Hyles** (then called **Celerio**) **lineata,** the white-lined sphinx moth. The moth has a tapered stream-lined body with delicate white, light brown, green, and pink markings. It has huge bulging eyes and a long tongue that, when not in use, is tucked neatly under the chin in a tight coil. The tufts of long scales at the end of the abdomen resemble a short tail. It flies fast, and willingly.

Adams and Heath enclosed thermocouples (fine wires for measuring temperature electronically) in the moths' thorax (the body part where legs and wings are attached). Then they used the long, thin wire leads from the thermocouples, which were attached to a temperature read-out meter, as a leash. They could take their moths out for a short morning flight down the halls of the zoology building and observe on an instrument dial if the thorax heated up. Flight is extremely rigorous exercise, which, as any runner knows, produces heat. As expected, the moths heated up in flight and, indeed, could not fly until the temperature in their thorax was close to that found at the end of their pre-flight shivering. This was because the hotter they were, the faster they could beat their wings, and they couldn't fly until their wings beat at a certain speed. The moths' thoracic temperature during flight was remarkably close to the normal body temperature of warm-blooded birds and mammals. But what was most surprising was not that thoracic tempera-ture was higher than air temperature, but that it remained at approxi-mately the **same** high temperature regardless of what air temperature they were flown at. The thoracic temperature of the sphinx moths during flight was regulated.

Regulation of body temperature is determined by Newton's law of cooling, which says that the greater the difference in temperature between a body and its surroundings, the faster heat will transfer between them. At around 68°F we feel comfortable, since the heat generated by our resting metabolism is just enough to keep our bodies stabilized at 98.6°. If it gets much colder than 68°, we must shiver to produce extra heat, much as a furnace kicks on when room temperature drops below some fixed point. Conversely, if it is very hot, we sweat and the blood vessels close to the skin dilate (we get red) in order to give off heat. Similarly, an air conditioner turns on when room temperatures rise above some fixed point.

If sphinx moths do maintain a steady thoracic temperature during flight, they could do so in one of two ways. If the heat produced from the flight activity was not enough, they would have to produce an increasing amount of extra heat when flying at decreasingly lower air temperatures. However, if the heat produced as a by-product of flight was always more than they needed to maintain a fixed body temperature, even at low air temperature, then they would have to actively lose more and more heat when flying at increasingly high air temperatures. Birds and mammals use the first mechanism to keep their body temperatures stable over broad ranges of air temperature, but since no insect had ever before been shown to regulate, it was not obvious which method they would use to keep thoracic temperature stable during flight: extra heat production or active heat loss. Adams and Heath concluded, "Large moths increase metabolism during active periods to offset heat loss and thereby maintain a relatively constant internal temperature. In this regard they may be considered endothermic, like birds and mammals." Thus they were concluding that the moths regulate by the heat-production mechanism.

But what actually did these researchers mean by "active periods" in a sphinx moth? A sphinx moth feeds by hovering. When it is not hovering or flying from one place to another, it is totally at rest. For someone who knew sphinx moths, "active" seemed to mean flight, especially since the measurements of metabolism or heat production were made "during periods of uniform activity following warm-up." Moths warm up for one *thing only, to fly. I could not visualize* **how** *the moths might increase their metabolism at low air temperatures in order to stabilize thoracic temperature, as Adams and Heath seemed to imply for flying moths. The upstroke and downstroke muscles of the wings in the thorax could be contracted simultaneously against each other for shivering, as in pre-flight warm-up. Or the muscles could contract alternately to move the wings up and down during flight. But they could not be used for warm-up and flight simultaneously. Did the moths beat their wings faster (to produce more heat) with decreasing air temperatures? The researchers made no mention that wingbeat frequency varied at all, implying instead that the muscle efficiency varied. But heat production of an exercising muscle cannot be simply turned on or off. If the* **moths** *could do it, it would be a new biological phenomenon.*

I had to think up other possible explanations. Heath and Adams had measured metabolism in small jars. I knew that if a moth is prevented from flight, it stops its wingbeat and cools immediately. After cooling it may again shiver to try to fly again. The lower the air temperature, the faster it would cool, and the more it would have to shiver to get ready for flight; the total metabolism averaged over a series of warm-up bouts would be higher. Were they really measuring warm-up rather than flight? On the other hand, maybe their conclusion was faulty and the moths were not regulating thoracic temperature at all — maybe they were not flown long enough following warm-up for thoracic temperature stabilization to have occurred. Or maybe if the moths did thermoregulate, they did it by getting rid of excess heat at the high air temperatures, rather than increasing heat production at the low air temperatures. Perhaps they did it by a combination of both heat production and heat loss.

I gave a seminar on my thoughts about the Heath and Adams paper to Bart's laboratory group. Bart accepted me as his student, and it was decided that "the moth problem" held more potentially interesting prospects than the other five thesis projects I had come up with. So I set out to investigate, first of all, whether or not sphinx moths can thermoregulate while they are in continuous flight.

It was a major problem simply to get sphinx moths. **Hyles,** *the moths*

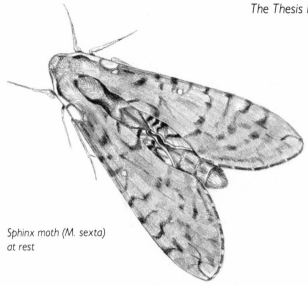

Sphinx moth (M. sexta)
at rest

that Heath and Adams worked with, occur in outbreaks for a few weeks or so, at some times in some places. Then they are often gone for years. It might not be a good animal to work with for a thesis problem, since I would need a steady supply of many. If a moth is punctured to take its temperature, it will be dead or damaged and cannot be used a second time. Insects are very delicate and usually short-lived.

And then I discovered **Manduca sexta**, the common garden-variety tomato "worm." These caterpillars could also be found all over the jimson weed out in the Mojave Desert. They also eat tobacco. The worms hatch out into beautiful moths with delicate gray and black markings and a brilliant patchwork of black and yellow on the abdomen. They were not only as beautiful as **Hyles**, but also bigger. I wanted large moths so that I could more easily insert thermocouples and other probes into them with a minimum of damage. The only problem was that the big moths mean big caterpillars. Each caterpillar ate about one fifteen-inch-tall green-house-grown tobacco plant per day, and I was rearing 50 – 100 moths at a time. It seemed for a while that I was becoming a tobacco farmer, although toward the end of my studies I discovered a synthetic diet of wheat germ, agar, sugar, and vitamins that the caterpillars thrived on quite readily.

Measurements of the thoracic temperature of moths that were either attached and flopping at the end of a short thermocouple lead or enclosed in small containers did not relate to the normal behavior of these animals. I wondered what the temperatures were like during **continuous** flight where the moths could fly as long and as fast as they chose. I considered implanting radio telemeters to monitor temperature, but that wouldn't work because the moths were too small to accommodate any known telemeter. Perhaps an infra-red spotting scope? No, that would only give me surface temperature of the moths' scales, and it wouldn't be accurate enough.

I had another idea. I would fly them suspended on a mill where they could fly in circles at any speed they chose, as I continually monitored thoracic temperature. The leads from the thermocouples would be strung through the hollow arm of the flight mill to a temperature recorder, where thoracic temperature could be continuously printed out on a chart. I designed the mill, rushed to the shop, and the thing was built. At the unveiling it sat on a little table in Bart's lab. A moth was put on. Graduate students gathered round. The temperature recorder started printing out the moth's temperature every three seconds. The chart paper started to roll, and the moth took off. The moth flew slowly at first, when its thoracic temperature was low. Gradually, from the flight exercise, it got hotter and hotter and flew faster and faster. The hotter it got, the faster it flew (and the more heat it produced and lost). After three to four minutes the rates of heat production and loss were equal, and thoracic temperature stabilized. My new toy worked. I would fly the moths at different air temperatures to see whether or not sphinx moths really did regulate their thoracic temperatures during flight.

Moth after moth went through its paces, and moth after moth stabilized its thoracic temperature, generating an excess body temperature of about 10°C above whatever air temperature I tried. That is, not one of these moths regulated its thoracic temperature. If air temperature was 15°C (58°F) they flew with a thoracic temperature near 25°C (77°F), and if air temperature was 25°C they flew with a thoracic temperature near 35°C. I was disappointed. These humdrum results suggested that there was no mysterious new physiological mechanism for me to decipher after all. Perhaps worse, it meant that I still had no thesis problem.

I had examined thoracic temperature by more sophisticated methods than Adams and Heath had, and my results contradicted theirs. Should I publish a paper and then move on to a more promising subject? Adams had written to me saying that they had a workable hypothesis for the mechanism of regulation in flight, as well as four graduate students working on the problem. They would very likely scoop me with the same results, so I would have nothing to show for my second year at UCLA. Maybe I should at least try to get into print before they did. Fortunately I didn't publish just then. More sophisticated methods do not necessarily make for better results, if there is a flaw in the ideas behind them. In this case, however, it wasn't so much a flaw as it was an important detail I had overlooked. And that detail would ultimately stand everyone's ideas on their head.

It was December, a time when it could be cool in the Mojave at night. I and a fellow graduate student, Gary Stiles, had camped in a wash among barrel cacti, boulders, ocatillo, and bushes of tuperow with blazing red flowers. The wash also contained the mauve green bushes of a mint with delicate blue flowers that were abuzz with honeybees throughout the day. The patches of tuperow attracted hummingbirds in the daytime. As the sun set and the coyotes howled on the distant ridges, we sat down to wait for sphinx moths.

I first heard a soft muffled fluttering. Then, looking like a small hummingbird, a **Hyles** sphinx moth hovered in front of me. It dipped into a flower, backed out, and went on to the next flower. In another instant the moth was fluttering wildly in my net. I grabbed it and stabbed it with the thermocouple probe: 42.5°C (108°F), or 32.5°C above air temperature! I remembered having seen **Hyles** flying near Los Angeles on a warm day, at 25°C. If the moth's body temperature is a simple passive phenomenon of the flight metabolism, as my flight mill experiments had suggested, then the Los Angeles moths should have heated up to 57.5°C (136°F). That was clearly impossible. They would have been cooked. There was something funny going on. They probably do thermoregulate in flight, at least when they are in the field. Why didn't the **Manduca** sphinx moths do it on my flight mill? Was it a species difference?

Back in the lab at UCLA I abandoned the flight mill. I simply allowed *the moths to fly freely in a temperature-controlled room. After they had been in continuous free flight for two minutes (to simulate activity under field conditions), I grabbed them out of the air and stabbed them with the thermocouple as I had done in the desert. The results were entirely different: now their thoracic temperatures were high, close to 40°C (104°F) and sometimes up to 43°C. Furthermore, the moths were hot even when air temperature varied from 15°C to 34°C. They were thermoregulating!*

Should I pick up the moths again and continue the study, or would I get beaten out in the race with Heath's lab, where students had been working on the moth problem for a long time? I had better find out what had already been done, so I wrote Heath about my results. He answered in December 1969 and told me that in five years of working with **Manduca sexta** *his lab had never recorded a moth's temperature higher than 40°C. Of course, he added, their animals were never "forced" to fly. Were my results screwy, then? "Forced?" All this time I had thought the point was to make sure the moths I took data from were in*

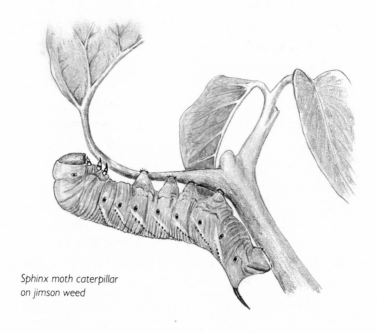

Sphinx moth caterpillar
on jimson weed

continuous flight. I felt some apprehension, but I knew I had been careful. Maybe they had missed something. I decided to continue with the problem, sensing that it involved some profound complexities that none of us was yet aware of.

As a first step in my decision to proceed, I spent a few months in the library reading about insect physiology in general and everything about sphinx moths in particular. Something in the known physiology and morphology might provide a clue. It would be necessary to collect more and more details on the problem until I could visualize it as closely as if it were a rock sitting in the palm of my hand. I wanted to find out how the moths were thermoregulating. Adams and Heath thought they did it by varying heat production, but their explanation didn't make sense to me. The heat produced by the moths during flight could all be an inevitable by-product of the flight metabolism. I needed to design an experiment that would separate out heat that might be produced for temperature regulation from heat unavoidably produced as an unwanted by-product of the flight metabolism. Upon some reflection I realized that one such experiment had already been done. The moths held aloft on my flight mill had a lower thoracic temperature (less heat production?) than those working harder in free flight. If the moths on the flight mill had produced heat specifically for temperature regulation, they should have been as hot as those in free flight.

Increases in temperature, however, are no proof of increased metabolism, or heat production. Metabolism had to be measured directly (most conveniently by determining the rate of oxygen consumption). Manduca sexta are reluctant flyers, and I had trouble getting them to fly nonstop in the huge pickle jar I used as a respirometer. But some did fly, and the moths had nearly the same oxygen consumption, wingbeat frequencies, and amplitudes at all air temperatures, as long as they remained in continuous free flight. When I suspended them from a tether (like the one on the flight mill), wingbeat amplitude and metabolic rate was immediately reduced by one half. They worked less to fly. This explained why they had lower thoracic temperatures than when in free flight; they produced half as much heat. Heat production did not vary at all with the moth's cooling rate or with air temperature, but only with the effort they made to stay aloft and flying. The inescapable conclusion was that they were thermoregulating in free flight by regulating the rate of heat loss, with heat production remaining constant. The moths on the flight mill did not thermoregulate simply because, with lower flight effort, they never produced enough heat to call for the dissipation response.

My conclusion was now almost precisely the opposite of what Heath and his co-workers had published. They said the mechanism of thoracic temperature stabilization involved the regulation of heat **production.** I said it was the regulation of heat **loss.** One of us was wrong. I could hardly claim to be right, however, unless I could prove a mechanism whereby heat is dissipated.

There are three ways vertebrate homeotherms may increase heat loss: by varying their insulation, by panting or sweating, or by increasing blood flow to the capillaries in the skin. But the moths could not vary their insulation like a bird or a hairy mammal. Their scales are fixed solidly in the hard exterior skeleton. Insects do not pant or sweat. They have only a single large circulatory vessel, the "heart" that traverses the abdomen and extends into the thorax as the "aorta." They do not have capillaries with which to direct blood to or away from the body surface. Furthermore, Norman Church, a student of Francis Wigglesworth, the father of insect physiology, had examined avenues of heat gain and loss in insects and had concluded that neither blood circulation nor evaporative water loss was a significant factor in an insect's heat balance. Whatever it was that the moths might do to thermoregulate, it would probably be a discovery.

I came across an obscure French paper of 1919 by Franck Brocher on the anatomy of the blood circulatory system of sphinx moths. The odd thing about these moths is that the aorta makes a loop through their thoracic muscles. In many or most other insects, it passes **underneath** these muscles before terminating in the head. Why should evolution introduce such a wrinkle unless there was some selective pressure for it? Brocher gave no speculations on the functional anatomy. But given my results I wondered if the selective pressure for the evolution of the loop *might have been heat dissipation — could the loop be a cooling coil for the muscles? In insects the blood is pumped from the abdomen into the thorax, and the abdomen of the moths remains relatively cool in flight. The blood entering the thorax from the abdomen must be cool, too, and common sense dictated that the anatomy of the sphinx moth was arranged so that cooling of the thorax would be enhanced during blood flow.*

My problem was now specific: Is heat loss regulated by way of the circulatory system? I couldn't stay away from the lab for more than a few hours at a time. Once I went to camp out for a weekend in the Sierras, but my mind was in the lab and I didn't see the birds or flowers. My measurements of heat production (oxygen consumption rate) and thoracic temperature showed that the moths thermoregulated by as much as tripling their rate of heat loss during flight at high air temperatures. But how could you measure heat loss facilitated by blood flow in a flying moth? Sphinx moths are extremely fast. You can't trail them in flight with instruments attached. I decided to mimic the overheating that normally occurs in flight and to control it myself.

Moths were tied down and heat was focused by a narrow beam of light from a heatlamp directly onto the thorax. Meanwhile, thermocouples in the thorax, as well as several placed in the abdomen, were used to record body temperature simultaneously with heart pulsations on a polygraph machine. Would the moth tell me the "truth"? The results made all the efforts and waiting worthwhile. The new pieces of the puzzle fit, and provided attachment places for other pieces.

A captive moth heated on its thorax with a focused beam of light stabilized thoracic temperature (at about 42°C, 108°F), while the temperature of the abdomen continued to increase! Further experiments showed that heat transfer into the abdomen (and thoracic temperature stabilization) was correlated with strong pumping action of the heart. This was clear proof that temperature stabilization of the thorax involved variations of heat transfer and not necessarily of heat production. My thesis advisers were not all that convinced, though.

Franz Englemann said that I needed to eliminate the circulatory system to prove my point. I had thought of that, too, but dismissed it as an impossibility. Now I thought about it again and had an idea: I would do heart surgery. I'd take a surgeon's needle, thread it with a fine human hair, loop it around the heart where it empties into the thorax, and tie the knot to stop the pumping action of the heart. This would eliminate all possibility of the blood's acting as a potential coolant. But would the moths with extirpated hearts fly at all?

They did indeed, but soon stopped with heat prostration at high air temperatures. Even air temperatures of 24°C were too high for them, as their thoracic temperature approached the lethal limit of 46°C (115°F). With their heat-dissipation mechanism totally gone, but with the insulating scales removed from the thorax, they could fly at air temperatures several degrees higher. This was frosting on the cake. To see what had never been seen before and to have it confirmed is an indescribable feeling. At this point I published a short report in Science to foreshadow a more detailed account where the full evidence could be given.

If I thought that the moth problem had been settled, however, I was sadly mistaken. I need not have worried about getting scooped. The other papers did come out on thermoregulation in **Manduca sexta**, but none of the workers had deciphered the mechanism of thermoregulation in flight. One of Heath's students showed that the moths' temperature sensitivity resides in the thoracic ganglion, adjacent to the thoracic muscles. Heath and his colleagues made numerous calculations to construct hypothetically derived relationships and models. Drawing from earlier work in the same lab, they now claimed (as before) that "measured rates of oxygen consumption in flying animals increase as air temperature decreases," causing them to conclude that flying moths thermoregulate in the manner of known endotherms generally measured at rest (by varying heat production). Their measured metabolic rates closely matched the calculated rates of how much heat the moths "should" produce.

Later I learned through correspondence that their flying moths had their wings amputated; the researchers assumed that wing-stub movements corresponded to actual flight. It was a detail that made all the difference, but they had not deemed it important enough to state in their published methods. I had already suspected that their supposedly flying moths had not been in flight at all; using their own data I was able to show that some of their moths "flying" at high air temperatures had rates of heat production identical to those of torpid moths at the same temperature. Perhaps these workers were blinded by their preconceptions based on higher vertebrate endotherms. It is easy to see what you are looking for, as any intent deerhunter can testify when he sees a big buck in every arrangement of brown ferns and branches.

After I left UCLA for the summer to go back to Maine, I stopped off at a symposium in Bloomington, Indiana, where Heath was scheduled to speak on the regulation of heat production in moths. It was a large new auditorium, well filled with people. Heath was a forceful speaker, and he presented his evidence for thermoregulation. Questions were pounding in my head: "What was your chart speed when you measured warm-up rates?" "Are you certain that your flying moths did not, even for a fraction of a second, stop beating their wings?" The questions had to be asked, and I put up my hand. The dean of comparative physiology, who was presiding, called on me, but hardly had I opened my mouth when he asked me to sit down. He suggested that I should see Heath afterwards.

So I did: "Look, we obviously have drastically different conceptions of this problem . . ." Our moth discussion was short. We wandered over to a room where a large group was having an informal get-together over beer. Heath talked about the hypothalamus of birds. There was no indication that he had any interest in discussing the moth problem. I was frustrated. One of the people in the group said to me, "I suspect that your moths and Heath's moths will, with time, become more and more alike." Perhaps, I thought. But how and by whom would that be decided? Which moth would the scientific community "see"? The experimental evidence had already been published. I couldn't provide better proof—if this didn't do it, what would?

It seems that it is not always the force of the experiment or the evidence that wins the day. It is also the force of the argument and prestige of the person presenting it. I would at least have to write a visible article to guide readers into the principles of insect thermoregulation so that they would want to delve into the details. I would have to offer a broad theoretical perspective. Only then could the results fit into a rational pattern and no longer be ignored simply out of convenience. With these thoughts in mind, I wrote a review article. After a rejection, I protested that the criticisms of the anonymous referee were not valid. The journal reconsidered, and the paper was published in 1974.

Six years later I overheard two students. Their conversation went something like this:

"Did you get that exam question on thermoregulation in big flying insects?"

"Sure. The insects stabilize thoracic temperature by shunting heat by the blood to the abdomen."

"How did they prove that?"

"Easy. They just tied off the blood vessel to the thorax."

"Oh! And what about the other question on . . . ?"

As in so many discoveries, any reasonably capable person could have performed the conclusive experiment. But what makes the experiment possible is the long struggle of ideas that leads up to it.

In a Patch of Fireweed

After four years in Los Angeles I returned to the farm. Would the realities overwhelm my expectations? We never did much farming there. We had rescued the house from collapsing, however. One of the two barns fell and rotted, and jewelweed flowers grew in its place. We had used one of the fields to grow a crop of potatoes the first year and then let it lay fallow. The field was invaded by airborne seeds of poplar, birch, and willow. Ruffed grouse and snowshoe hare were beginning to move in behind them.

After a shed had partially decayed and fallen, we set fire to the remains. The fire smoldered long in the humus, killing millions of buried seeds. The ground became bare and sterile, and as after the Mount Saint Helens volcanic eruption in Washington, the first plant colonizers were fireweeds. The fireweed seeds, dangling on delicate silky parasols, came in on the wind. They could have come from the colony along Wilson stream in East Wilton or from Newfoundland. Nobody knows where they come from, but fireweeds presumably send out probes for colonization every summer. Here in Maine only those rare seeds that happen to land on a patch of burned and sterile soil will sprout and survive. In other spots the entrenched natives always beat them out. Only the toughest immigrants, whether plant or animal, can compete on ground that has already been colonized.

The slender green fireweed shoots sprouted from the soot. Aphids, also blown by the wind, landed on the shoots and tapped the rainwater they had taken up from the ground, as well as the food energy these plants had made using energy tapped from the sun. The aphids multiplied. A small black ant found the aphid colony and informed its nest mates. The ants used the aphids' surgary secretions and, in turn, tried to guard their "livestock" from predators. But then a small red and black ladybug flying by detected the aphids and laid her eggs near them. Then the beetle and her larvae ate the aphids.

The fireweed kept growing. When several feet tall, the plants sprouted brilliant scarlet inflorescences that shone like flames above the green patch. The flowers attracted sphinx moths at night. In the daytime they were visited by the red admiral butterfly, by blues, hairstreaks, tiger swallowtails, and others. Solitary metallic-green bees took pollen, and the ruby-throated hummingbird hovered to take nectar with its tongue, and most common of all were the furry bumblebees foraging for both pollen and nectar.

My thesis hunt was over. It was a relief not to feel urgent pressure to solve a problem, to be able to observe without needing an excuse for it. I fished for sunfish and yellow perch in Pease Pond. But my unfocused relaxation didn't last long, after I noticed the many bumblebees at that patch of fireweed. Were they hot-bodied like the sphinx moths? I started taking their body temperatures.

It was fun to chase bumblebees in back of the barn among the fireweed flowers — far more fun than chasing insects in the confines of a temperature-controlled room. And the initial results were similar: at least in moderate air temperatures at midday the bumblebees had high thoracic temperatures. But did they thermoregulate while foraging? In order to get body temperatures at the lowest air temperatures, I would have to catch them at dawn. Would the bees foraging at sunrise be as hot as those at high noon?

Every day I went at dawn to the patch collecting body temperatures throughout the day. The bees came at 10°C (50°F) at dawn, as well as at much higher air temperatures. I plotted the thoracic temperature data on a graph, and I could see that the bees maintained thoracic temperatures at 30–35°C (86–95°F) at air temperatures from 10°C to 25°C. So the moths had not been unique. The bumblebees were also thermoregulating, in some way.

On a research project I usually try to graph my data on the same day I collect them. From day to day the points on the graph tell me about my progress. It's like a fox pursuing a hare. The graph is the hare's track, and I must stay close to that hare. I have to be able to react and change course frequently. Also, since nature is complex I let it lead me, trying not to get too far ahead, so that I don't have to backtrack.

I was stalking a new and interesting animal, the thermoregulating bumblebee. The bees stopped flight every few seconds as they perched on flowers to pick up the nectar and pollen. They might be maintaining their elevated thoracic temperature by shivering during the short stops at the flowers. It was a difficult idea to test because the stops at flowers were so frequent and so short, often exceeding twenty per minute. The stops were short because bees could extract all of the nectar with one quick jab of the tongue. I decided to manipulate the bees' behavior by making them stop longer — long enough for them to cool. Then I would be able to separate heat left over from flight metabolism and additional heat produced by shivering specifically for temperature regulation. I enriched some of the fireweed blossoms with drops of sugar syrup. A bee that found one of these enriched flowers should stay until it had taken all the sugar, and if the solution were very thick and viscous, then the bee might lap away for a long time. It should cool off precipitously within several seconds of landing, unless it shivered to counteract passive cooling the instant it stopped flight.

As expected, the bees sometimes stayed for several minutes at flowers with the sugar syrup. I could measure thoracic temperatures after they had stopped flight for varying lengths of time. Thoracic temperatures were independent of perching durations. Indeed, bees who stopped flying because they were detained by ample rewards often maintained higher thoracic temperatures than those who spent more than half their time in flight. To confirm my hypothesis that the bees who

stopped flight on flowers still produced heat specifically to keep themselves warm, I needed a nonshivering bee. How does one keep a bee from
shivering to see what a lack of shivering does? Well, dead bees don't
shiver. I heated a dead bee to 32°C (90°F) at an air temperature of
10°C (50°F), and this bee started to cool at a rate of about 15°C per
minute. There was no longer any doubt: foraging bees produced heat
specifically to keep themselves warm at low air temperatures. Sphinx
moths, which do not stop flight while foraging, did not have the same
need. Indeed, it was this major difference from sphinx moths that initially
sparked my interest in bumblebees.

One thing always leads to another. On cool mornings or cool days while
I searched for bumblebees, I noticed that they might be found on
jewelweed, fireweed, and milkweed, but they didn't appear on goldenrod or aster until the weather was warm. Why should the bees discriminate flowers on the basis of air temperature? Maybe goldenrod and aster
don't secrete nectar in the morning. It is true that some flowers don't
secrete their nectar until afternoon or until the temperature is high, but
nectar could accumulate over days and be available in sufficient amounts

With Phil on the Penobscot

to satisfy the bees at all times. So what was the answer to the bee's discrimination? A partial answer came during a fishing trip.

My old friend Phil Potter and I lifted the Grumman canoe onto the special canoe racks he had invented for his car, and we took off. Phil remembered to take the "fly dope" (a bottle of Scotch) for internal consumption. I remembered to take along my thermocouple thermometer.

By dawn we had passed Norridgewock and Skowhegan, well on our way along the Piscataquis River toward Greenville. We were soon out of the towns and away from fields, taking the dirt road to Kokadjo. Now we were into the North Woods, which stretch unbroken for another 160 miles, as the crow flies. Kokadjo is only a small clearing in the woods. The gatekeeper of the paper company's logging road let us in, and we traveled another forty miles on dirt roads past Chesuncook, then to the west branch of the Penobscot River. The fish would be really big here.

As we got out of the car and stretched, I saw bumblebees on the goldenrods near the stream. They didn't seem to be moving very fast. Why weren't they in a hurry, like those back home? It seemed cold for bumblebees to be out on goldenrod, I thought. But fields of clover were at least fifty miles away, and I didn't see other flowers they might use instead.

"Just a minute, Phil, let me check these bees out." They were **Bombus terricola,** and air temperature was 13°C (55°F). I had not seen them foraging from goldenrod at this low temperature at the farm, and so I took some measurements. The thoracic temperatures varied, and I could tell that the data did not fit my graphs back home. Curiously, many of the bees appeared unable to fly.

Phil was unloading the canoe. "You coming?"

"Just a minute!" A big cloud was moving across the sky. I had to wait — this could be an interesting natural experiment. What would the bees do when the air temperature drops? Phil could see I was becoming preoccupied. Having unloaded the gear, he took out the fly dope.

Phil had grown up in the North Woods. When he was sixteen he worked during the winter at lumber camps in northern Maine, hewing out cedar railroad ties with an ax. Later he worked in the woolen mills, primarily as a loom fixer. I had spent a few years doing chores on his farm, earning money to buy my first bolt-action rifle when I was sixteen. He also invited me to stay at the farm with him and his wife Myrtle during some school vacations when my parents were away on expeditions. Phil and Myrtle and I had explored quite a few streams, ponds, and rivers in the North Woods, catching our share of brook trout and shooting white-tailed bucks in the fall. I knew he didn't mind putting off the fishing right now. He seemed to enjoy sitting back to watch me in action.

When the cloud passed over the sun, all the bumblebees stopped flying. The thoracic temperatures of all the bees I measured were less than 30°C (86°F), the minimum level needed to fly. Some of them had a thoracic temperature only one degree above air temperature. These by themselves were trivial details. But in terms of a larger picture they hinted at exciting possibilities. I began to see the high energy costs of thermoregulation as well as the benefits bought with it. The costs of keeping warm in the cold would increase, and the benefits would not exceed the cost, since goldenrod flowers have little nectar. If bees have a foraging strategy based on economic costs and payoffs, then a lot of the puzzling results would fit into a clear pattern. Maybe the bees were allowing themselves to cool off because they could no longer afford the heating bill in terms of calories gained from these meager nectar supplies.

A few hours later, after other clouds had come and gone, Phil wasn't feeling the mosquito and blackfly bites; nor did he seem concerned about catching "the big ones." We loaded the canoe and the gear and drove back. We couldn't stop to fry fish as on other trips, but bought some hamburgers in Skowhegan. It was a good fishing trip. I took home a big one. I didn't know it then, but it would keep growing with each year.

What I took home was the answer to why bumblebees in some areas did not forage from goldenrod early in the morning when it was cold and why they were cold-blooded when they did. A few calculations of potential calories expended and collected, and the balance between energy investment and reward, lent support to my idea that the bees balance their energy budget and respond by way of their foraging and *thermoregulatory behavior. These ideas were not original, since Bart and his colleagues at UCLA had known this all along for vertebrate animals. It was intriguing to me, however, that what applied to some birds and mammals might also apply to a bee.*

Bees seemed able to calculate profits and loss and to behave appropriately. When flowers with ample nectar are available (fireweed, jewelweed, milkweed), the bees tend to stay away from the less-rewarding flowers, especially at low air temperatures where they have to expend more calories to keep warm than they can get back from the flowers. But they can afford to forage from low nectar producers at higher air temperatures, when the energy investment for foraging is less, because less energy needs to be invested for thermoregulation. In the Kokadjo area they had no choice — if they were going to forage at all at low temperatures, they were forced to visit the low-reward flowers. Since goldenrod blooms as a dense panicle with thousands of individual florets, however, the bees did not need to maintain a continuously elevated thoracic temperature, since they could restrict foraging — at least for a time — to the same panicle. They could walk all over it and let their flight muscles cool. By this line of reasoning, they should also let thoracic temperature decline on high-rewarding flowers where it takes them a long time to fill up, such as on the fireweed blossoms I had enriched. But this response has not evolved, presumably because there are no flowers in nature that have nectar rewards large enough to detain the bees for more than a few seconds.

The summer ended all too soon, and I returned to UCLA that fall for postdoctoral work in Bart's lab. I was working on the physiology of pre-flight warm-up in sphinx moths, but I wasn't exactly sure how to proceed further. Should I look in greater detail at the coordination of neurons, at the presumed temperature set-points, or at something else? Should I become a "pure" physiologist, searching out underlying neural connections, or an "environmental" or comparative physiologist, dealing also with the adaptation and evolution of different species in their present environments? Not having a ready answer to this question, I was easily swayed by circumstances.

I had barely started my work when Franz Engelmann (one of my thesis advisers) told me about a job opening for an insect physiologist at Berkeley. I wasn't sure whether to apply. Returning to the University of Maine had always been in the back of my mind. Also, I did not consider myself an insect physiologist, perhaps because I came from a vertebrate laboratory and my work on insects was rather limited to my thesis problem. Franz convinced me that I'd be a fool to pass up this opportunity, since it was primarily a research appointment. To be able to spend stretches of unobstructed time on my own research was too good to ignore, so I went to interview for the job. I liked the people, the campus, and the whole set-up more than I had anticipated, and they liked my seminar on thermoregulation in sphinx moths. I was hired and went to Berkeley as an insect physiologist.

The first year I didn't have to do any teaching. I was to set up my lab and get my research program started. There had just been another big budget squeeze, and the promised equipment could not be provided yet. The lab had been stripped of such useful items as glass tubing, stoppers, vials, and corks. But drawer after drawer was filled to capacity with weird bulbs, vacuum tubes, resistors, and capacitors. The shelves were stacked with instruments that were army surplus of one sort or another, and I didn't know what they were or what they might be used for.

My first concern was to write up for publication the results of my summer field research in Maine. I spent a lot of time in the library, sitting in the coffeehouses in Berkeley, thinking idly, scribbling, and thinking some more. This was a great privilege that can, in our culture, only be provided at a university. The greatness of a university (and of a culture, for that matter), it seems to me, can be measured by the opportunity it allows individuals to go their own way to specialize and perfect their individual aptitudes and interests, whether it be research, teaching, musical composition, or painting. I was hired to do insect physiology, but I was left free to speculate, research, and write about bees and flowers.

I was impressed by one curious fact: the behavior and physiology of the bees were responsive to the nectar rewards of the flowers. I began to wonder whether the bees were, in an evolutionary sense, being manipulated by the plants. Obviously, the nectar in flowers evolved for one thing only: to ensure pollination. The sugar in the nectar was the reward or currency used to manipulate the pollinators into working for the plants. The amount of nectar would then, of course, also be a product of natural selection. Too little nectar, and the pollinators would choose another plant and not visit the flowers at all. Too much reward at one plant, on the other hand, would be just as bad, or worse, because it would cost the plant a lot of calories to produce the flowers and nectar; then the pollinator would stay at the plant, become satiated, and thus not go on to spread the pollen to other plants. The optimal amount of reward depended on a delicate balance that could be swayed by innumerable variables. How, and in what amounts, should the plants dispense the nectar to ensure the most pollination at the least cost? Much depended, first of all, on the foraging costs and capabilities of the pollinators. How much energy are they able or willing to spend?

My theoretical premise suggested an explanation for a dazzling array of seemingly unrelated facts. Flowers had evolved scents and pretty colors to lure insects, their intermediaries for mating. But they could not accept these proxy matings indiscriminately. It would not pay a plant to have its pollen collected and deposited on another flower where the pollen would do no good. The flowers of different species must be distinctive enought to be recognized by the pollinators. Did this explain, as a first approximation, why there is usually such a variety of flowers in any one bog or mountain meadow? Did it explain why the bog blueberries that the huge queens visited in the early spring had delicate white bell-like flowers while the rhododendron that bloomed at the same time with much pollen had showy purple flowers instead? It might all depend on what other species of plants were available through evolutionary time, when they flowered, what they offered, and who pollinated them.

The complex questions to be asked seemed endless. As I kept reading everything I could find on pollination, I found that many bits and pieces of the individual questions had been asked already. However, they had not been put together into a coherent theory indicating that the movements of pollinators were manipulated on the basis of energetics. The concept of energetics in pollination ecology seemed a highly useful one to me. It might help to explain the behavior of bees and other pollinators, as well as the flowering ecology of their food plants.

Still, it was all related to the evolutionary past, and I could not do experiments on the effects of various evolutionary scenarios to prove or disprove my ideas. It is in the spirit of science to be all-knowing whenever possible, but nature is not always willing to cooperate. One can always guess. But sometimes more progress can be made by probing for secrets when, for one reason or another, the time is right for data to be collected. Science must yield to rationalization, but rationalization must be based on relevant data. Just then the time was right for me to get data on the bumblebees' energy balance and their interactions with flowers. I had stepped into a patch of fireweed and come out a bumblebee economist.

Capitalist Bees

My work on bumblebee foraging and energetics was first published in the Journal of Comparative Physiology, *a technical journal. The thought occurred that it should go out to a wider scientific audience, and so in 1973 I wrote "The Energetics of the Bumblebee" for* Scientific American. *Shortly after it appeared I received a letter from Herbert Mitgang of* The New York Times *asking for an even less technical essay for the opinion-editorial page. He thought there were interesting conclusions for people to be drawn from the bee's use of energy and nature.*

Now, I wasn't ready to extropolate to men solely from bumblebees because the conclusions to be drawn about energy, economics, and survival were universal, at least from a biological perspective. A biological view of the world is in many ways an economic one, but with important differences from what an economist might see. An ecologist sees the world in terms of a much larger community — one including other organisms with energy flow or economic transactions involving costs and payoffs at different points along the spectrum. The economist, in contrast, is more apt to isolate man out of the community of which he is a part. From such a viewpoint, where pure profit is seen apart from the rest of the energy flow, an exploitative relationship to nature is encouraged if it can bring quick payoff at low cost. But in nature almost every part feeds back on itself, and the costs are sometimes hidden and almost always delayed. The larger and more complex the web of interacting parts, the longer the delay in the feedback. There are usually many simultaneous feedback loops, all with different time delays.

In bees the exploitative action on flowers results first and foremost in a positive feedback to them — the fulfilling of their economic needs or wants. But, quite commonly, directly on the heels of this exploitation is a negative effect — the depletion of immediate food resources. The third feedback, however, is a positive and less immediate one — the pollination of flowers or an increase in future food resources. The fourth feedback, occurring over immense evolutionary time spans, is also positive, as the flowers become adapted to and make rewards available to the bees. The bees, though, act only as if they were on the first short-term loop; they exploit for their own benefit, not that of future generations. Would the "bee story" be attractive to the public because it saw only the long-term positive effect and judged short-term exploitation in terms of that? Or did the public think that the bees had somehow learned to live in harmony by forgoing exploitation of their environment? No matter what the answer, it was a generalization, and generalizations can be dangerous because they tend to gloss over details.

A very important detail for any comparison between human and bee societies is that bees use only renewable energy resources, economizing drastically when resources become scarce. To them, the "energy crisis" is a way of life. Bees are almost always in a crunch because they convert all of their food and energy resources into offspring, which grow up to need more energy, and so on. Not endowed with foresight, they do not curb their reproduction and hence always precipitate an energy crisis for themselves, even if by some miracle they could be suddenly awash with gallons of high-octane nectar. Nothing new here. What more could I say, from my patch of fireweed in Maine, to readers sipping their coffee and munching their bagels in the high-rise apartments of Manhattan? I tried.

The Times *printed my article. There it was in black and white, "The Energy Crisis: A Biological Vantage Point," complete with a cartoon of a monstrous giant (a capitalist?) rolling barrels of oil that gushed black and horrible across a beautiful landscape. The picture implied, "Ah — if only* we could be like those beautiful bees living in harmony with the flowers!" I guess the message didn't get across because I meant to say that we are, like the bees, exploitative without regard to long-term consequences. That is the problem. If we see harmony, it could be from our perspective of seeing only the consequence of the long-term feedback from the perspective of the short.

Bees live in constant struggle, and the flowers ultimately control them because the bees cannot take more than the flowers offer. They dole out energy resources in minute amounts, and so the bees have become exploiters par excellence. As far as they are concerned, the pollination of flowers is incidental. And because the plants have gained a measure of control through evolution, the more the bees exploit their environment, the flowers, the more they promote the reproduction of plants. With us, it is the reverse—the more we exploit our environment, the more impoverization can result, especially if we depend on nonrenewable resources like coal, oil, and uranium, which of course have no stake in our survival.

I wrote another piece for the **Times**, "What Bees and Flowers Know," to give the earlier one another perspective. Bees and flowers are symbiotic, but the paradox is that they exploit one another. We have similar symbioses with our crop plants and domesticated animals. Indeed, all organisms live interdependently. It is essential in each of these relationships that the opposite member remains populous so that it cannot be obliterated. Corn and chickens, as unique organisms, would probably become extinct if people stopped wanting to eat them. Each takes, each gives, each needs the other. If one partner succeeds in taking too much, then the other is decimated and can give no longer. Wolves take from the caribou, but if they take too much they will be temporarily fat and eventually starving. If we build up our human population to the limit on the energy base of coal and oil or nuclear energy, we are exceeding our present ecological bounds and thus destroying our partnership with plants that we will have to depend on later when the oil and coal run out. Maybe having more to use at this time will only postpone the crunch—an even tighter crunch—until later. Thinking in terms of *decades, in terms of the short feedback loops only, may be short-sighted. We are, right now, only reacting to circumstances, and that will make us prisoners of circumstances in the long run. Such views are quite common among biologists, who almost necessarily think of complex relationships in the context of community and in terms of time.*

Some might call my speculating on "What Bees and Flowers Know" as gross anthropomorphizing. It is, however, only a comparison; attributions of human characteristics to animals, or vice versa, are not implied. The problem comes with assumptions of volition, if, for example, I were to maintain that bees forage for more than their daily needs because they are conscious that they need to store up honey for a cold and rainy day. But there is no need to explain bees' behavior in that way, any more than one would need to say that the heart is conscious that it pumps blood in order to oxygenate the body tissues. It is legitimate to say the heart pumps in order to circulate blood, since that is its evolutionary objective, although we know it actually pumps blood, in the proximal sense, simply because it contracts and expands. Biologists automatically separate proximate causes (what immediate factors cause the bee to go out to forage) — from ultimate ones (why they have evolved to forage as they do). When I talk about bees and flowers I talk about mechanisms, not intent. Presumably the bees' economy operates without any forethought or intent on their part, while our economic systems are often controlled.

One thing always leads to another, as even my own publishing history shows. Paul Landon, managing editor of **Business and Society Review,** *read my article in* **The New York Times** *and wrote to ask if I would do a short article along the same lines, on "The Limits on Adam Smith's Theory of Greed." I didn't know much about Adam Smith, but there seemed to be something stimulating there to think about.*

Adam Smith's long book, published in 1776 and titled **An Inquiry into the Nature and Causes of the Wealth of Nations,** *is a treatise not only on economics, philosophy, and sociology, but also on nature in the larger sense. Smith described what he saw, and he thought about a community of people interacting over time. But what he observed as the way things* *are came to be interpreted as the way things should be — a philosophy, in other words. To me, though,* Wealth of Nations *provided some ideas to test on my bumblebees.*

Smith maintained that every individual in society attempts to improve his or her own well-being, and by the constant action of each and all to better themselves, there is competition and society as a whole evolves and prospers. The baker, he says, does not bake bread to feed society. He bakes bread in order to sell it so that he might exchange his services (with money) for those of the carpenter, the farmer, the teacher, and so on. Because of the competition among bakers, society ultimately receives the best bread that can be made at the time.

Smith seemed to be describing not only society but also some aspects of biological evolution. His economic model is a "natural" system because it reduces conscious control to a minimum. Isn't that precisely how the bee's economic system operates? Previous analogies between bee and human societies had been drawn along "communistic" lines, ignoring competition. It was almost as if the bees in a hive sit around on a warm summer night and, over a glass of mead or two, decide among themselves which flower each of them should forage from in order to promote the good of the hive. An alternate model from the opposite end of the political spectrum might hold that the queen is a dictator or manager who directs whatever has to be done in the hive and where and when her workers should forage.

When I returned to Maine in the summer, I discovered a bumblebee colony in an old torn mattress in our barn. One bee after another landed on the mattress and crawled into the hole along the ripped edge. Some carried yellow pollen lumps attached to the two little "baskets" on their last pair of legs. Others carried orange, green, brown, white, or multicolored loads. Others brought no pollen at all, though their abdomens were distended with nectar. Clearly, different bees had gone to different kinds of flowers. But how had the division of labor been decided?

My field experiments had already suggested that each bee in a colony exercised considerable independence of the others. Each seemed to do its own foraging at the best flowers it could find. There was an echo here of the Adam Smith model, with the specialization of each individual based on individual initiative. Might this system result in an efficiency that fed back to the colony as a whole?

I determined to test the questions directly by watching individual bees. I could mark individuals (with little numbered tags glued onto the thorax) right after they emerged from their cocoon, and then I could follow these marked bees in the field, where there are many flowers to choose from. In practical terms, however, this seemed impossible. Bees fly very fast and far. How could I possibly follow a bee through its course in the field? Another idea came. In order to follow the foraging careers of a number of bees, I would have to use a small colony, and the foraging field (the bee's world) would have to be small enough so that the bees stayed within my sight when they went out to hunt. I decided to make an enclosure stocked with different kinds of flowers where I could keep track of individual bees.

With timber and poles I built a frame and enclosed a portion of meadow, filled with different flowers, in the back of our barn. The framework was then covered with wire screening, and in two days of steady work, my "field" was ready to receive bees. I took a portion of the bumblebee hive out of the mattress, housed it in a wooden box with a round entrance hole that I could open or plug up with a cork from a wine bottle, and set the box in my field. I let out one bee, closed the hive, and then followed the bee until it returned to the entrance to be let in. Usually when I let one bee in, another was waiting to leave, and then I followed that one. And so it went — from dawn till dusk, every day for most of the summer. There is no stopping until the end, unless you decide to waste your investment in the experiment.

To get significant research results probably requires having a good nose for good problems, much as a hound can sniff for a spoor of a strange animal in the woods. But I wondered if ultimately it might not boil down to tenacity of pursuit. I persisted in the chase because, if I didn't get it done in one summer, I would have to duplicate a lot of effort to get the experiment going again the next summer. To be only partway in, to stop halfway in a race because of fatigue or boredom, is not to be in the

Bumblebee collecting pollen from wild rose

race at all. No hawk ever caught a bird by giving less than a full effort to the chase. If it did, it would totally waste all that effort and it would not long survive.

As Adam Smith would have predicted from his theory on human society, each newly emerged bee on its first trip out into the world, in search of a profession, tried its tongue on a number of different kinds of flowers. It might probe into aster, goldenrod, or spiraea flowers, and then try out jewelweed or turtlehead. It made many handling mistakes at the complex, more inaccessible flowers, which were also the ones that contained the greatest rewards. After a bee had finally mastered manipulating a flower of complex shape so that it got all of the nectar, it visited only a few others. As more and more bees learned to visit the high-reward flowers, the competition decreased the nectar rewards at these flowers. Eventually when the jewelweed, say, was overloaded with bees, the new recruits in the field found more rewards at other flowers *and specialized in them instead. I could not listen in to what the bees might be saying to each other in the hive, but my observations were consistent with the notion that they said nothing at all. They sorted themselves into the different flower professions by individual initiative. As a consequence, the nectar and pollen coming into the colony was at the highest possible rate. And because no flower was long left unexploited, the hive collectively responded quickly to changing conditions.*

The bumblebees starting out their foraging careers were flexible and open to change. They were not fully committed to one kind of flower, until they happened to come upon a flower with relatively rich rewards. If rewards were not too great, they would, in addition to their major specialty, also have a minor flower specialty. The minor could quickly become their major if rewards in the major dried up. Individuals were constantly comparing. However, the older they got, the less able they were to switch their profession. For example, when the jewelweed became scarce, I observed some old jewelweed specialists who refused to switch to the very common aster, even though other bees of the colony utilized asters all the time. Not only did these bees conform to Adam Smith's model of competition for optimum functioning of society; they also seemed to conform to some human weaknesses, or vice versa.

I went back to Berkeley that fall excited about my observations. After I wrote up my findings, I was finally ready to think about "The Limits on Adam Smith's Theory of Greed" for the **Business and Society Review.** *My feeling was that Smith's economics described the bumblebees' foraging economy quite well, and the system obviously worked. The major difference was that our medium of exchange for labor is money, and that of the bees is honey and pollen. Also, instead of Smith's society of butchers, bakers, and candlestick makers, I saw a society of pollen foragers on rose, nectar specialists on jewelweed, and nectar foragers on turtlehead or aster. The labors of each contributed directly to the bees' society (actually a family) in that all the honey went into the same pot. The distribution of wealth was communal, but the acquisition of wealth was not. In Smith's model, the distribution of wealth is through the medium of direct exchange. The distribution of wealth in the bee society* is by exchange only indirectly, since a bee deposits and withdraws from a community pot.

Smith said that exchange is not possible without the accumulation of capital; exchange gives freedom to individuals because it permits them to specialize where they want to, or for what they are best suited in terms of their needs and abilities. My previous studies had shown that different bees specialize on different flowers. In human society, there is also extensive specialization. When the carpenter buys a cabbage from the farmer, he is exchanging the labor of making cabinets for that of growing cabbages. Without exchange of labor through money, both the baker and the farmer would have to grow their own cabbage, bake their own bread, tan their own leather, make their own shoes, weave their own cloth. If they had to do all of these tasks, they would probably be inefficient at most of them. Efficiency is only possible through specialization, which relies on exchange. Exchange also ensures that only products needed or wanted by society will be produced. Unneeded services are soon eliminated by the "invisible hand" that regulates the system.

The spirit or intent of Smith's system was to encourage freedom. The world now has factories that produce pollutants and set restrictions on otherwise common resources. It is no longer just butchers, bakers, and cabinetmakers. This is where the "limits" come in, in order to preserve the freedoms we all value. So I wrote about limits for the **Review**.

All freedoms are subject to constraints, and reality (constraints) keeps changing. The circumstances must be taken into account. Smith said, "In the race for wealth, for honors, and preferments, [every man] may run as hard as he can, and strain every nerve and muscle, in order to outstrip all competitors. But if he should jostle, or throw down any of them, the indulgence of the spectators is entirely at an end." In other words, according to Smith's laissez-faire, it would be unthinkable for a factory owner to dump toxic wastes into the water because he must pay society for the privilege of doing so, no matter how long the feedback loop until the negative results — the cost of production — are felt. The manufacturer, not society, must pay for the costs of production of what he sells. "Greed" would be the invisible hand acting for the good of all. There was *nothing in Smith's philosophy that allowed one person to brutalize another by force, fumes, or other interference with personal liberties in the name of freedom. To the contrary, there is nothing in the spirit of competition that would allow the government to give away a tract of wilderness in Alaska that belongs to you and to me, unless the public should agree on the price. Fifty years ago the wilderness had no "value." Like the air, it was free for all to take. Now the laws of supply and demand have dictated a price. It is a precious commodity, owned by all of us, and cannot be given to a sheep rancher, or a mining company, without a fair price. Most important, the wilderness is also the property of people yet unborn. How can one establish its price?*

In the world of biology, and of society, no one formula provides all the answers. There is no $e = mc^2$ in the biological world of animals and the social world of people. Reality in the natural world is too complex to be dealt with in terms of absolutes — thus any system or model is potentially dangerous, if taken too seriously. It is more important to know what it is that we are trying to accomplish. If it is freedom and personal well-being, then the same laissez-faire policy, with appropriate costs and payoffs for the various alternatives, might be as appropriate now as in Adam Smith's day.

And how would Adam Smith see our world? He felt that the role of government was both to protect people and to give competition free reign. Our government standards specify building taller smokestacks (so that toxic chemicals will be carried to distant, rather than close, neighbors). Smith would have specified, instead, that the effluents be routed back through the front door of the factory whence they came. In this way laissez-faire competition between factories would produce the best way of dealing with the toxins. It would be survival of the fittest, with the invisible hand of competition protecting us all.

Honeybee Swarms

On warm, clear Indian summer days in the early 1950s, Floyd Adams, Robert Goodwin, and the boys, Jimmy and Billy and myself, would scamper over the back fields through the goldenrod, bee box in hand. Or we would sit among the asters in the apple orchard waiting for the arrival of returning bees. This was all a part of bee lining. It seemed like one of the occult arts as the grownups debated whether it was better to use a box with one or two compartments, whether it needed to have the natural bark on the outside, whether you needed to scent the box with drops of anise. They had strong ideas about what kind of tree the bees were likely to be in and how far a bee tree would be away from water. I felt very ignorant because I couldn't understand the basis for any of their assertions. Nevertheless, the grownups allowed us a hands-on experience.

The bee-lining operations started simply enough. I held the bee box in one hand and the cover in the other. I would approach a honeybee feeding on an aster or goldenrod and place the box under the bee and then slap on the cover, trapping the bee. The captive buzzed in the box for some minutes as it attempted to escape. Eventually it would find the sugar syrup we had put into the honeycomb on the bottom, stop buzzing, and begin to feed. At this point I removed the cover and placed the box on top of a stake with a small platform on it, which I had driven into the ground in an open spot away from trees. We then sat and waited for the bee to emerge from the box.

After filling up with syrup, the bee left the box, flew rapidly back and forth downwind from it, made several large circles and then, when it was but a speck in the distance, flew directly toward its hive. We waited patiently. Floyd looked at his watch, and we timed the bee's trip to its hive and back. That time, plus the number of recruits the bee brought back, gave us some way to estimate distance to the hive. After many recruits had arrived, we slapped the cover back onto the box full of feeding bees and released them at another field near the hive. Occasionally some bees returned. We would then have a "crossline," and from it we could get a rough idea where the bee tree was located.

Finding the tree in the dense woods wasn't easy because you had to extrapolate to an imaginary point in the forest, and even if you found it, the bees might be hard to see if the hive entrance was high up in the tree. A day or so after finding the bee tree we would troop through the woods, laden with axes, a crosscut saw, a smoker, many pots, and perhaps a few bottles of home brew for the grownups to quench their thirst.

After the tree was split open, we lifted out the combs and shook off the bees. The air was full of thousands of them, all searching for their hive in the air above our heads where there was no longer a standing tree. We placed the combs with brood (eggs, larvae, and pupae) and pollen into the frames of a new hive we had brought to house our bees, and we tied the combs securely with string. The combs with honey filled our pots. The new hive, set onto the split empty tree with the remains of the bees' nest, soon had bees at its entrance standing with their abdomens held up in the air, fanning their wings. They were emitting a scent that attracted their nestmates, who then began streaming into the hive.

After a week the wild bees had cemented their combs into the new hive and taken in the spilled honey. We went back to the hive after dark, plugged the entrance, and brought it home to set up in back of the house. I spent a lot of time lying in the grass in front of the hive. One after another bee with distended abdomens and bright yellow lumps of pollen on their hind legs streamed in. It seemed that there was a strange intelligence at work within the hive. The thousands of inhabitants were storing food for the future, as if they knew what they were doing. Of course it was all supposed to be "instinct," which I didn't understand. But I crept closer to my hive, satisfied to watch the bees come back with their loads of yellow pollen.

Later, when I was sixteen, my father gave me a book for Christmas, **Bees: Their Vision, Chemical Senses and Language** by Karl von Frisch. The text of this book, first published in 1950, was scarcely a hundred pages long, and it was written so clearly that even I could understand it then. It described von Frisch's experiments leading to discoveries on the mechanism of communication, and his book made the bees seem even more fantastic to me. They were fantastic to him as well: "No competent scientist ought to believe [those discoveries] on first hearing." I believed everything, though, and was sure this was the last word that would ever be written about bees.

Eleven years after this book was published, Martin Lindauer, Professor von Frisch's most distinguished student and co-worker, also published a splendid, plainly written study called **Communication Among Social Bees**. I was especially impressed with Lindauer's observations and his work with swarms.

The swarming of honeybees usually occurs in late spring. The old queen and several thousand of her daughters leave the hive abruptly; another queen and some residents stay on. In a few hours or days, the swarm establishes a home of its own in a new place. In the interlude the swarm coalesces into a beardlike mass on a tree branch or other firm object. There it remains immobile, until some signal stimulates all of the bees to leave in one buzzing cloud and fly directly to a cavity where the new home will be established.

Martin Lindauer studied honeybee swarms after World War II when he was working at the Munich Zoological Institute. He found that most of the swarms he watched chose cavities in the ruins that littered the city. From his careful studies he learned to predict the direction and distance a given swarm would fly before it left. He once startled an innkeeper by telling him he had come not to drink his wine, but to observe the arrival of a swarm of bees. Sure enough, they came and entered a hole in the wall above the speechless man.

Lindauer had observed that some of the bees, called scout bees, left the swarm cluster to investigate nest sites. They came back, often covered with dust, and began to dance on the surface of the clustered swarm, apparently indicating distance and direction of potential hive sites. Conflict arose when several scouts simultaneously indicated different cavities. In cases like this, which occurred routinely, the bees still eventually left all together, usually going to the best place among all those the scouts had located. How did the bees come to a consensus about the best future home?

Before performing its dance, a scout visited a cavity several times. After it had started dancing it would continue visiting the site occasionally, but it would stop advertising if the cavity became too hot or too wet or undesirable in some other way. The scouts were also attentive to one another's dances, visiting the other potential home sites indicated. They were not only able to change their "minds," they also were converted, if they found another scout's nest site better than their own. Eventually all the scouts repeatedly visited and advertised the same site.

At this stage the question becomes how the swarm is aroused to leave as a unit and how it is guided to the new nest site. Several researchers have observed that, a half hour or so before the swarm lifts off, the scouts make rapid buzzing runs over and into the swarm. Perhaps this makes the bees get ready, but takeoff is sudden and some other signal is probably involved.

Once a swarm has taken flight, the 10,000 to 30,000 bees fly in a cylindrical formation some fifteen yards in diameter. A swarm may take off even if the queen bee is held captive, but it soon returns if she is not with them. Presumably swarming bees must smell the queen substance (9-oxodec-2-enoic acid) in order to keep going. The queen, though, does not guide the swarm. Apparently that is done by "streakers"—bees that fly much faster than the group. The streakers dart repeatedly through the swarm in the direction of the new hive. Once the bees are at the new hive, the scout bees indicate the entrance by releasing the scent from their Nassanoff glands at the tip of the abdomen. In a few minutes the thousands of bees stream into the cavity that will be their new home.

After learning of the fascinating communication system of the bees, deciphered by von Frisch, Lindauer, and their students, I was all the more awed by these insects. Usually I kept a hive or two, just to have the amazing creatures nearby. When I was teaching at Berkeley I kept an observation hive near a window in my lab on the second floor. I used it to have a constant supply of animals for experiments on the bees' thermoregulatory physiology and also as a teaching tool for classes.

In the spring the honeybees from my observation hive swarmed, which could be a nuisance. Sometimes they took up temporary residence on the cars in the parking lot below. The many cars seemed to confuse them. They milled around for a long time, apparently trying to decide which one to settle on or in. The swarms also settled in the olive tree next to the entrance of the entomology building. They were "my" bees, and I was under some pressure to remove them. So I would climb into the low tree, snip off the branch the bees were hanging from, and drop them into a cardboard box and tape it shut.

What could one do with a swarm of bees "in the hand"? It seemed that the opportunity to have a captive swarm practically in the laboratory should not be thrown away. There might be an interesting puzzle to solve. Bees keep their hives heated for incubation of their brood. If the hive is not warm enough, pupae die or emerge with developmental defects such as deformed wings. I knew, however, that the temperature is not rigidly controlled in the portions of the hive that do not contain the brood, and so thermoregulation was thought to be primarily for the benefit of the immature bees. A swarm consists entirely of adult bees without eggs, larvae, and pupae, suggesting that swarm thermoregulation may not exist. I recalled an old published report by Anton Büdel. He found that the temperature in the core of a swarm was high, near $35-36°C$ (97°F), like that in the brood nest of hives, but the outside layer or mantle of the swarm was relatively cool. It seemed odd that only **some** of the bees in the swarm should be warm. Honeybees are social animals. How was it decided who keeps warm and who freezes?

The first problem was that of handling swarms. My research career in honeybee swarms had already been cut off abruptly years earlier at UCLA while I was still fumbling around for a thesis problem. I had brought a swarm into the laboratory in a wooden box. The moisture emitted by the bees over the weekend warped the box, creating a crack along the side through which the bees escaped. Worse, the room in which the box was housed had a narrow gap between the door and the floor. To everyone's horror there were bees flying all over the halls on three floors of the zoology building on Monday morning. After that I had shied away from working on honeybee thermoregulation. This time, at Berkeley, I devised a plexiglass container. Plexiglass absorbs little moisture. My swarm cage could be readily assembled and disassembled, and the bees could be easily put in or taken out. It was a huge tube (with a diameter of 16 inches) that rested on a flat floor. Another flat piece of plexiglass drilled full of small holes was used as a lid. The holes would serve for ventilation, as a foothold for the bees (enabling them to form their swarm under the lid), and as openings for the insertion of measuring devices. When I wanted to measure respiration I could seal a second lid of solid plexiglass over the perforated lid. Air from the sealed system could then be circulated through an oxygen analyzer to determine the rate of energy expenditure (heat production).

I captured many swarms by simply shaking them from the branches into a cardboard box, and in the laboratory I would shake the bees out like beans from a sack into my plexiglass vessel. A few pieces of string dangling from the lid served as a ladder for the bees to climb up and reform their swarm cluster on the bottom of the lid. I could then carefully lift the lid with its attached swarm and probe to take temperatures.

As Büdel's preliminary temperature measurements of a swarm had indicated, most of the bees in swarms were relatively cool. Indeed, they were not warm enough to be able to take off immediately in flight, and this explained why it was so easy to capture them and to shake them from one container into another.

To answer the first question of whether or not swarm temperature is regulated, I subjected swarms to air tempratures ranging from 0°C to near 30°C in a large environmental chamber, probing them with a thermocouple attached to the end of a knitting needle with a cm rule etched onto the side. I could thus get temperature measurements and know the specific locations within the swarm where the measurements were taken. These measurements showed that not all of the bees are cold. Even when it got colder, the center of the swarm maintained its temperature near 35–36°C, or it increased slightly. Odd—I had not expected an increase. The outside, or mantle, temperature was also regulated; although it stayed cool, it did not drop below 15°C when air temperature was lowered to near zero.

I wondered what would happen when a swarm decided to take off in flight. The bees on the outside of the swarms I measured were much too cool to fly. A swarm would not take off in the lab, since no scouts could leave and search for nest sites. Even if a swarm could take off, I didn't really care to be around to take measurements.

But there was a solution. I implanted thermocouples in a swarm, and since the length of the thermocouple leads was variable, I put the swarm, with thermocouples inside it, into the great outdoors. The wire leads from the thermocouples could be extended to plug into a chart recorder by my desk in the laboratory. I could simply look up from my desk and check the chart to see what the temperature was at any particular time, past or present, in any portion of the swarm. Even if I wasn't watching the swarm, there would be a record of swarm temperature at takeoff. This certainly beat ramming a knitting needle into a live swarm, and it was more informative because the swarm was left undisturbed in its natural habitat. As expected, scouts soon issued from the swarm, returned and danced on the surface, and eventually the swarm left. The charts showed that a half hour or so before takeoff, the mantle bees heated up to the same temperature as the core bees. How might the changes in swarm temperature be regulated?

Our body temperature is controlled by the hypothalamic portion of the brain. This part of the brain receives information from sensors in the skin and core by neural pathways, and it sends out commands by other neural pathways to the body periphery and the muscles, to counteract the challenge of the cold. A swarm does not have an interconnecting system of neurons. How were the core and peripheral temperatures regulated? The swarm is composed of individual bees, and in honeybees individuals are coordinated by a complex communication system. Did thermoregulation also include a communication system, involving scents, sounds, and special bee messengers?

All of the available literature on thermoregulation in honeybee hives said the bees act like a "superorganism" with the bees in the core attempting to increase heat production as it gets colder outside, so that the rate of heat flow to the periphery increases sufficiently to keep the outside bees warm. Was this model an accurate one and, if so, would it apply also to swarms outside the hive? If the model was really true, then it seemed to me that the bees in the core would have to know what the temperature was at the periphery. The core bees are subjected to an unvarying temperature, and hence, if they did not have information on what bees on the outside mantle were experiencing, they would not be able to respond.

I designed several experiments to test if or how bees on the outside of the swarm might "talk" to core bees. In order to see if sound communication was involved, I recorded the buzzing on the edge of the swarm at low and high temperatures, allowed a swarm to form around a microphone, and played back the noises the bees made on the periphery to those at the center of the swarm. No response. I even played them the sounds of African "killer" bees attacking a microphone I had recorded in Kenya on a previous study of honeybees, and the music of Bach and Beethoven. The swarm was not affected by the muffled sounds of angry buzzing or by Bach. Perhaps the bees on the outside were talking, but those on the inside weren't listening. Were they signaling each other with scents instead? To find out I connected a tube between a swarm kept at room temperature to another kept in the cold room. Air with its attendant scents was pumped from one swarm into the other, but again swarm temperatures remained unchanged. The bees apparently weren't wafting chemical signals for temperature control either. I also separated a swarm into two parts, one with and one without a queen.

Honeybee with
regurgitated fluid

The temperatures of the two swarms remained the same, indicating that there was no central directive coordinating the swarm's response to changing temperatures.

Were there messenger bees running physically between the core and the periphery, carrying the information directly? That could be tested by preventing bees from traveling back and forth. I suspended a light screen sock that was open at the bottom from the lid of the respirometer vessel. Bees shaken into the respirometer crawled up and filled the inside of the sock and, on the outside, surrounded it. Now the core was physically isolated from the mantle — but temperatures within this swarm were identical with those of other swarms. Furthermore, I marked bees on the swarm mantle with paint and found that the bees on the mantle tended to stay there, especially at low air temperatures, even when they had an opportunity to move. Apparently the bees of the core and mantle do not tell one another about local temperature by way of messenger bees. The model of central directives as seen in humans and other vertebrate animals clearly did not apply. Something along different lines must be going on. I didn't know where to begin to look for an answer.

When in doubt it is best to do something, anything, rather than nothing. I measured the metabolic rate of swarms at different air temperatures. The metabolic rate of swarms that had settled and were at rest was surprisingly low. It was not elevated above the average *resting metabolism (at that temperature) of individual bees. Of course the bees in the center had higher resting metabolism because they were hotter, but there was no extra energy expenditure for heat production. There was only a small increase above average resting metabolism at the very low and the high air temperatures. Moreover, in swarms of from 1,800 to 16,000 bees, where the passive rate of heat loss (determined from "swarms" of dead artificially heated bees) varied by a factor of eight, there was little tendency for small groups to have a higher metabolic rate than large ones. In fact, I recorded some of the highest metabolic rates per given weight of bees in large swarms. So the rate of heat production wasn't geared primarily to counteract heat loss.*

Furthermore, since a large swarm has a greater proportion of its bees at the core than a small one does, and since the higher the body temperature the higher the resting metabolic rate, the observed metabolic rates could be understood in terms of the **passive** *heat production of heated bees, rather than in terms of elevated heat production to keep warm. The original hypothesis that the core bees increase their metabolism to keep the mantle bees from getting too cold was now turned on its head. Bees in the core of a swarm should have no problem keeping themselves warm. Instead, they might have a problem getting rid of excess heat that is produced merely as a byproduct of their resting metabolism. They are heated, apparently, because the bees crowding around them prevent the escape of heat, causing higher body temperatures, which cause the production of still more heat.*

The pertinent problem now seemed to be how bees in the core avoid overheating. I tried to answer this first by analyzing the response of the entire swarm and then observing the behavior of the individual bees. One obvious swarm response was that, as air temperature was increased, the swarm greatly expanded in both length and width; it became more porous and increased its surface area for passive heat loss. In addition, the bees on the mantle became more widely spaced and their heads pointed outward; at lower air temperature they bunched tightly together with their heads pushed like shingles into the mass of bees, a behavior much like the huge aggregations of emperor penguins in their Antartic rookeries during a snowstorm. The much looser high-temperature configuration, on the other hand, facilitated the flow of air through the mantle to the centrally located animals, and this air removed heat from the interior.

Watching swarms maintained in the container through the transparent lid to which the bees were attached, I could also look directly into the swarm. At high air temperatures, stationary bees hung in curtains, and there were passageways between these curtains where the bees moved freely. The temperature within the passageways was lower and more variable than in the surroundings; the passageways served as ventilatory ducts. As air temperature was lowered, bees from the periphery crawled inside, filling up the passageways, and the swarm became a solid mass of bees with no ventilatory ducts to the outside. Clearly the mantle bees, who were experiencing the external air temperature, responded by huddling more closely in the cold, and this had the secondary effect of helping to stabilize the core temperature. It seemed that the mantle bees were concerned primarily with their own temperatures.

The data were consistent with the hypothesis that the mantle bees not only huddle but also shiver at low air temperature; this is to maintain and defend the lower limit of individual body temperature, near 15°C (60°F). In so doing, however, they reduce the temperature difference for heat flow from the core, and this helps to keep the core temperature high. Nevertheless, there is only a limited amount of heat to go around. This applies not only to honeybees but also to the bumblebee, keeping its abdomen cool so that its thorax stays hot, and it also applies to humans. Menopausal hot flashes in women were recently studied by Fredi Kronenberg and her co-workers in the department of obstetrics and gynecology at Columbia. (Kronenberg went to Columbia to work on humans because of her expertise in thermoregulatory physiology. Her Stanford thesis, appropriately enough, was on thermoregulation in honeybees inside the hive.) Kronenberg observed that, during a menopausal hot flash, peripheral skin temperature (measured in the finger) increased on the average by 4°C in four minutes. Meanwhile, body temperature fell by about 0.5°C as heat was diverted from the body core.

In the bee swarm, however, without a central directive to coordinate temperature control, all of the bees were acting independently, but not all of them had the same capabilities and requirements. I reread Büdel's paper and learned that the bees in the swarm mantle, who tend to be older workers, are foragers. These bees are best able to shiver and to tolerate lower temperatures. The young "house" bees, on the other hand, who are less able to shiver, tend to seek the warmer temperatures inside the swarm. These facts fit my model in which the core bees are heated passively and the mantle bees shiver to generate heat in order to keep from being chilled below 15°C, the minimum body temperature below which they lose motor control and fall from the cluster. By not shivering until they are cooled to 15°C, the bees of the mantle prolong the swarm's limited reserves of food. Yet the swarm's thermoregulation makes possible a quick response in a wide range of weather conditions at a very critical stage in the colony's cycle.

The wonderful ability of the honeybee swarm to regulate its temperature economically is apparently the result of each bee's doing its own thing. I wondered why I had not thought of it sooner.

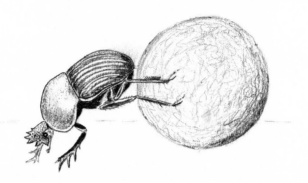

African Dung Beetles

The rainy season was well under way in Kenya. The red Tsavo clay was already thoroughly moistened. It sprouted grass, which provided the economic basis for a vast and varied spectrum of life. Within days the plains were a pea-green sea undulating in the gentle tropical winds. Red termite mounds projected from it like the prows of sunken ships, and some that were worn down more than others served as lookout points for handsome impala bucks defending their harems. The bleached bones of elephants that had died in the red dust of the drought were disappearing in the rising tide of new life. Giraffes stood lazily under the scattered acacia trees, reaching up to nibble the emerging leaves. And the birds were everywhere. The quelia finches — transformed physiologically at the mere sight of the green grass — had started to sing, to build nests, and to lay eggs and care for young. The male widow birds had changed their drab sparrow plumage to one of long black feathers, and they twittered as they shook their plumes to show off to their brown females. Grasshoppers and termites fed on the grass, and birds fed on the insects and the grass seed.

In the distance we could see the rounded backs of elephants projecting like smooth boulders above the plain. As we came closer we saw them move ponderously through the green, occasionally flapping their monstrous ears. Still closer we could see them leisurely but steadily pulling up one swath of grass after another with their waving trunks.

We had come to study elephant dung beetles, and my immediate task was to collect a few mounds of elephant dung to fill a large plastic garbage bag. I didn't dare venture very far onto the plain. We kept our distance, waiting eagerly for some elephant to defecate. Since they were eating continuously, it was inevitable that one would, but they were in no hurry.

We found a spot where the herd had passed, and there we collected the precious beetle bait. Secure with our cargo, we drove a short way off the clayey road into the scrub. From the trunk of our rented car I scooped out two handfuls of dung. As we waited for the approaching darkness we heard the doves in their evening chirring. A hornbill called one last time. Soon the onslaught of the dung beetles would begin. Our primary objective was to measure the body temperatures of flying beetles as they arrived.

George Bartholomew and I had received a grant from the National Geographic Society to study dung beetles, and I was happy to be reunited in the field with Bart, my former major professor at UCLA. In the years since finishing my work on the physiology of thermoregulation in sphinx moths, I had examined physiological mechanisms of thermoregulation in other insects and had developed a theory on the evolution of temperature regulation that might apply to all insects, and perhaps to other animals as well. Our work with the beetles was designed to examine this theory.

The central idea was that those few animals who, at some time in their evolutionary history, experienced selective pressure for rapid and continuous locomotion (such as running or flying), and at the same time were big enough to retain heat produced from their own activity, would evolve to regulate a high body temperature. Dung beetles exist in a large range of sizes, from tiny insects smaller than a house fly to creatures bigger than sparrows. Selective pressure to fly far and fast in search of dung should put a premium on the ability to elevate metabolism or power output. The increased metabolism for activity, in turn, would unavoidably cause increased heat production as a by-product, and then there would be a selective pressure to tolerate, and function at, the increased body temperature so created. Small beetles, which should not be able to heat up no matter how much heat they produced, would not have experienced the selective pressure to tolerate and function at a high body temperature, though they might be just as active.

Not only would very large beetles have to tolerate higher body temperatures, they might also have evolved mechanisms to dissipate heat. The higher the body temperature they could tolerate, the less heat they would need to dissipate. But the higher the body temperature they evolved for flight, the more energy they must invest in pre-flight warm-up, and to reduce metabolic costs of pre-flight warm-up they must evolve mechanisms of heat dissipation. Thus those insects that have evolved to dissipate heat during flight would also be the ones likely to produce it before flight.

The biggest beetles we were hoping to catch at the elephant dung were the giant **Heliocopris**. I remembered catching them in mist nets we set for birds on Mount Meru in Tanganyika many years before. Invariably after elephants had come by at night, these insects were found entangled in the nets in the morning. **Heliocopris** dung beetles have thick legs with backward projecting spines that allow them to get traction like the treads of a caterpillar tractor or a tank. And they have immense power to push forward. The front of their head and the sides of their forelegs have knifelike extensions that function as digging blades. Normally their great strength is used to sink shafts into the hard soil beneath an elephant dung pile. They are nature's analog of a backhoe, though one that flies. Would they come tonight? If so, would they be hot-blooded, as we expected?

We can never know all of the selective pressures that a beetle of any one species has experienced throughout its evolutionary history. We can only make a few educated guesses. A small beetle of a few milligrams presumably cannot heat up significantly by its own metabolism no matter how vigorously it flies, since it does not have the mass to store heat in significant amounts. It would, however, still be under the same selective pressures as the others to fly fast and vigorously. Could the small beetles, of the many other species that we also expected to come to the dung, fly at a low body temperature?

The beetles came with the dusk, at first sparsely and singly. I captured them in an insect net or grabbed them the moment they landed and quickly took their temperatures. Bart recorded air and body temperatures in a notebook, rolled the animals up in paper, and numbered them so that we could identify them later. We would weigh them that night, and the next morning after breakfast we'd have the satisfaction of plotting the data on a graph and maybe seeing something we could not see before. After taking only a couple of dozen measurements, though, a general pattern was already emerging: not one of the tiny flying beetles had an appreciably elevated thoracic temperature, but all of the large beetles were hot-bodied. Things were falling into the pattern we had predicted.

Completely unexpected, however, was the spectacle at the dung pile shortly after nightfall. As it got darker and the guinea fowl and doves became silent, there arose a sonorous humming and buzzing as the air became filled with thousands — perhaps hundreds of thousands — of beetles. They came flying upwind toward the dung in the dense brush, and hundreds of them were descending on the elephant droppings each minute. Within an hour the dung pile was a seething mat that contained more beetles than dung.

Why was this dung such an attractive food? Did it contain some special flavor? Pondering this, I threw away the remnants of a not very good ham sandwich we had packed at the safari lodge. Within seconds beetles were leaving the dung and eating the sandwich instead, and within a few more minutes it was covered with a mass of beetles who seemed even more lively and eager than those on the dung. Maybe the sandwich wasn't so bad after all.

The thousands of beetles (most of them very small ones) now busying themselves in the dung pile were apparently eating as much and as fast as they could. Heliocopris came also, sounding like small helicopters as they approached.

The Heliocopris beetles started digging shafts into the soil directly beneath the dung. Males and females formed pairs at the dung pile and cooperated in digging. When the shaft is complete, the female stays at the bottom. The male grabs armfuls of dung from underneath the pile and carries them partway down the shaft, where he dumps them on the head of his waiting mate, who carries it down the rest of the way. Three feet or more at the bottom of the shaft, the female excavates a chamber where she fashions the dung into several pear-shaped balls that are stacked narrow end up. She will deposit one egg in the top of each ball, and the developing larva will consume the ball from the inside, eventually eating all but the outer shell. The female stays down in the ground with her young as the clay above her later becomes solid during the dry season. Eventually she dies there. Meanwhile, a year later, the rains come again and soften the soil, a signal for the now fully developed offspring to escape upward out of the chamber where they have been safely sealed away from almost all predators and the harsh physical environment.

Some other beetles fashion the dung into round balls on the surface and then roll the balls away to eat them in privacy or to bury them to feed larvae. Beetles not much larger than the end of one's thumb made dungballs as large as a baseball and nearly as round. As soon as they had finished making a ball, they got behind it and started pushing and rolling it with their hind legs, doing handstands and walking backward on their front legs.

We were impressed by the speed of the dungball rollers who worked at night. We had previously observed rollers in the daytime, and these insects moved at a leisurely pace. Perhaps they didn't need to hurry then because there was so little beetle activity at the dung. The beetles working at night faced more competitors, though fewer predators. The nocturnal rollers were moving so fast that the patting movements of their front legs on their balls became a blur in the scrutinizing beam of our flashlight. After they had formed a ball, they rushed to roll it away. Balls were moving as rapidly as lightly tossed tennis balls in all directions from the dung pile. But there were also a few nocturnal slowpokes, and we wondered if they had low body temperatures.

We laid out a level racetrack free of sticks and other debris, where we could measure the distance beetles rolled the balls as we timed them with a stopwatch. As expected, ball-rolling velocity was very much related to thoracic temperature. Beetles with a relatively low thoracic temperature of 28°C (82°F) rolled balls on the average of 2 inches in one second, while those with a thoracic temperature of 43°C (110°F) rolled them four times faster, or 8 inches in one second.

Rolling per se did not necessarily result in any measurable increase in thoracic temperature, even in the largest beetles. That is, they must have been heated by another activity. The contractions of the small muscles moving the legs did little to heat the thorax. Apparently the massive thoracic flight muscles in hot running beetles shiver, heating the whole thorax, including the muscles that move the legs. The hotter these muscles became, the faster the beetles could run.

There was good reason for the nocturnal rollers to move quickly and to be hot-bodied. If they weren't fast, they couldn't make and get away with a dungball before the other beetles at the pile had eaten or buried all the dung. There was also another advantage in being hot-bodied. Newly arriving dungball-rolling beetles first ran all over the pile apparently looking for others who had already made a ball. If they found another beetle with a ball in its possession, they immediately attacked and tried to take the ball away. No beetle gave up its ball without a fight. The contestants, both of whom attached themselves to the dungball with their hind legs, faced each other and tried to flick each other off with vigorous upward thrusts of the forelegs. The loser of the fight was usually thrown several inches, often landing on its back. Before it could right itself, the winner in great and frenzied haste rolled away the ball, if it had not already been torn apart during the contest.

We soon ran out of dungballs to use in our experiments because they were destroyed in the fights or because they were being consumed by the hundreds of tiny beetles boring into them. We then tried to make surrogate balls by mixing the clayey soil with our drinking water and coating these clay balls in dung. In their great haste for a bargain, the beetles accepted these cheap imitations. A beetle seemed to become even less discriminating, and accepted a false dungball eagerly, when another already owned it. We gave two beetles the same ball at the same time, and a fight ensued that lasted until one emerged the winner. Almost invariably the winner of the contest was the more hot-blooded,

In combat over one of
our clay balls

even if it had a size disadvantage. Generally, larger beetles had a fighting advantage over smaller ones, but small beetles could make up for their disadvantage by raising their muscle temperature. Boxers also warm up before a match.

Even as we were becoming progressively more absorbed in the excitement at the dung pile, we were not totally unaware of our surroundings. We had to remember that we were in Tsavo Park, in Kenya, East Africa, where tourists were forbidden to leave their cars even in the daytime because of the potential danger from wild animals. No sane person would walk out in the bush at night with only a flashlight. The darkness had enveloped us. Occasionally lions roared. The sound of a roaring lion is the most frightening, yet most beautiful and exciting, sound I've ever heard. The volume seems unreal, as it rolls across the bush country like thunder and is answered by other lions in the distance. Was the roaring of some of the lions driving prey toward other members of the pride lying silently in ambush? I kept alert for the snapping of a twig in the nearby brush and nervously shone my flashlight there, half expecting to see a huge set of eyes shining back like headlamps. But whenever we turned on the flashlights, plumes of dung beetles flew at us along the beam of light. It was a strange sensation to be hit in the night by hundreds of beetles that crawled into our hair, ears, and eyes, and that had previously been crawling elsewhere. There were incentives both to turn on the lights and to keep them off. It was a matter of delicate tradeoffs.

The night grew darker still, and the lions stopped roaring. The dung we had carted out had long been rolled away, buried, or eaten in situ. Only dry chaff and coarse, short remnants of grass stems remained of what had been a juicy substance of solid consistency. The swarms of beetles dwindled.

In the trunk of our car we had more dung in reserve, and I spread it out on a flat patch of ground cleared of grass. Within minutes the buzzing crescendo resumed. We had run out of ham sandwiches, and also water, but by now were so absorbed in collecting data to fill our graphs that we thought of little else. The beetles were behaving in all sorts of ways that made it increasingly difficult to leave them. They were becoming more interesting than the elephants themselves. Indeed, I felt that maybe elephant dung ought to be touted as a tourist attraction for the wonderful spectacle it provides.

We observed, as had long been known, that the rollers make dungballs for their young to eat as well as for themselves. Quite often, when a male beetle made a ball, a female beetle of the same species jumped on top of the ball and hung on tightly. She made little buzzing noises, perhaps signifying to the male that she was a potential mate. The male then rolled the ball away, with female attached, and buried it. After the ball was buried the female nibbled on it, and while she was distracted the male mated with her. Dungballs made by males and offered to females for the apparent purpose of winning a mating privilege are called nuptial balls. In some species the females does not ride on the nuptial ball, but follows the male closely as he rolls it along. In still others she appears to assist the male in rolling the ball. In many species the females make brood balls specifically to rear larvae in. The larvae of all ball-rolling beetles grow up feeding on the buried dung into which the females oviposit.

Dung is the sole food, both for the larvae and the adults, of some two thousand species of beetles in Africa. They are the clean-up specialists of the plains, with enormous practical value. The good work of the dung beetles had not gone unnoticed by economic entomologists. Cow pats can be a real problem in a dry climate. If unattended they harden and take valuable land out of production, one splat at a time. Millions of acres were going to dung annually in Australia, where there were no large herbivores until man brought cows. The dung was not only covering the land, but was also a breeding place for bloodsucking flies that reduced milk production in the cows they attacked. The pesky face flies could also irritate humans to distraction. Then dung beetles were brought to the rescue from Africa. Such beetles are now also being introduced into the southwestern United States. It is estimated that in California alone they are worth about $5 million annually just in terms of increased milk production in cows. In Africa the beetles make and maintain the soil that supports the huge populations of herbivores that graze on the plains.

One night, as on many others, we quit work on the beetles when we were exhausted, feeling satisfied and happy. We sank down into the seats of the car, started the motor, and pulled out onto the road. A night plover with huge eyes ran through the beam of our headlights. We came to the first bend in the road, and there was a lioness stretched out across the road. She looked at us unblinkingly as we stopped, and only after a long and deliberate stare did she slowly get up and let us pass. Her presence, too, was tied in with the beetles'.

Whirligigs

The biological station of the University of Minnesota sits among forests at the northern end of the state along the edge of Lake Itasca. A small sandy stream drains this great lake, and a plaque indicates that it is the headwaters of the Mississippi River. The shallow water gurgling here over the rocks and among the weeds will make its way to the Gulf of Mexico.

The lake is a fisherman's paradise. Blue herons stand on slender legs waiting for minnows in the shallow coves, and kingfishers dive for the fish from dead snags. Ospreys plunge into the water for walleyes from high in the air, and submerged loons chase fish quietly under the surface. Bald eagles pick up dead suckers along the shore and harry the ospreys to drop their catch in the air. Humans troll for walleyes from canoes and motorboats. Some fish eat other fish; most of them eat insects.

I toss a slightly stunned horsefly onto the lake from the boat dock of the station. The fly is captured by the surface film of the water. In its struggles it buzzes and broadcasts concentric ripples. The ripples make light and dark rings on the shallow bottom as the water bends the light rays. There is a swirl in the water as a largemouth bass comes to the surface, and the fly disappears in a splash, in the way that most insects do who fall into this lake.

To one side of the dock, at the edge of the wild rice, there is not one insect but a group of hundreds, perhaps thousands. Each of them is as large as a horsefly. They sit quietly on the water surface, and no fish seem to disturb them. They are whirligig beetles (Gyrinidae). The beetles are so closely crowded together into a raft, or pod, that they often touch. There are several rafts close together making up an aggregation. The next aggregation of rafts is about a mile farther down the shoreline.

The beetles in a raft mill around slowly, showing no indication of going anywhere. Indeed, I've seen several rafts of beetles in almost the same area near the dock for several days in a row, and the beetles of these rafts only seem to move when a canoe comes close to them. I ponder why they are in a group, when they feed, and why they aren't eaten by the fish all around them.

Gyrinid beetles are known to be predators and scavengers. Living at the interface between the aerial and aquatic worlds, they collect insect prey from both worlds. The water surface is the launching place where many biting flies as well as mayflies, stoneflies, and dragonflies leave their aquatic nursery to embark on their adult lives. Later they fly down to the water surface to lay their eggs, and after that they fall back in, drown, and are eaten. In addition, the water surface also collects victims from the air and land who have never lived in or on the water. Beetles, flies, moths, grasshoppers, and other insects fall onto the water and are captured by the surface film. Even the most agile and strong aerial acrobats can be rendered helpless if they fall onto the water, and after some awkward and futile struggling they disappear beneath the surface with a whirl and a splash of scaly fins. But the water beetles live on the water surface all of their adult lives.

Whirligig beetles are nature's analog of an all-terrain vehicle. When they need to disperse from one body of water to another, they turn into good flyers and they can also walk on land. Their forte, nevertheless, is existence on and in the water. They can submerge and swim under-water, but more commonly they skim across the surface. A whirligig's body is smooth and shaped like a teardrop to minimize drag and is lubricated by oils. A beetle's body displaces only half its weight of water since the other half is supported by the surface film, which is depressed like a dimple.

*Naval architects calculate the maximum (economical) velocity of conventional ships by the formula: (maximum velocity in knots)^2 = 1.96 (shiplength in feet). Applying this equation to a whirligig beetle (length approximately .45 inches) yields a maximum speed of 0.28 knots. But the beetles do much better, tirelessly swimming up to 2 knots. The paradox is related to scale and to the different kinds of waves generated by a beetle and a boat. The large waves behind the boat are called gravity waves because, once made, they are influenced primarily by gravitational forces. Those of the beetle, on the other hand, are much smaller capillary waves, which are influenced not only by gravity but even more by surface tension forces. Both waves travel, as do light and sound waves, at rates determined by their size, but the speed of the capillary waves **increases** as they get smaller, while the gravity waves **decrease** in speed the smaller they get.*

How does this relate to propulsion energetics? To greatly simplify the matter, it is energetically most economical to travel exactly in the trough between two waves, that is, at the speed characteristic of a wave having a wavelength the same as or larger than that of the length of the ship. As a ship moves faster, the gravity waves it creates become longer, and when wavelength has increased to approximately ship length, the ship has one wave at the bow and one at the stern. The amount of energy required to travel faster — that is, to push over the top of the waves created — is too costly for commercial vessels, and the same seems to apply to the surface-dwelling gyrinid beetles. The beetles move forward following the innermost wave as it expands outward. They situate themselves between a bow and a stern wave, but since these waves are capillary waves that are extremely rapid at the short wavelengths of approximately the length of the beetle, the beetles swim much faster than predicted by the formula used by naval architects.

The beetles would not be able to realize their top economical speed, however, if it were not for their efficient hydromechanical design and their superb propulsion devices. The beetles' two pair of hind legs kicks at 50 – 60 times per second, slightly faster than the wingbeat frequency of many hummingbirds and sphinx moths. The middle pair of legs moves

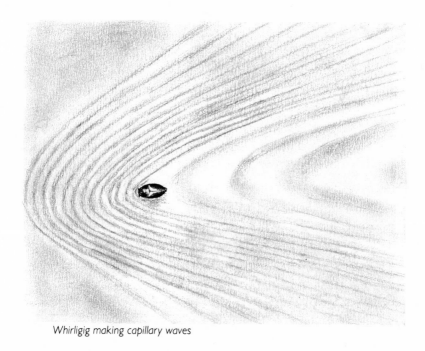

Whirligig making capillary waves

at half the frequency of the hind pair, while the front legs are folded tightly against the body, kept in reserve for grappling hooks.

The beetles are the fastest of all water insects. Yet despite their water-skimming capabilities, here were hundreds of beetles lolling in a group and hardly moving at all. They were not feeding, mating, or doing anything else either, and they appeared to stay in the same area for days on end. Odd. I tossed a stick into their midst. Beetles took off wildly in all directions, making the water boil, but within seconds one or several rafts of beetles were convening again. The original aggregation was reformed in less than a minute, as if some kind of mysterious force were pulling them back together.

I had expected the beetles to be dispersed as much as possible, for this should be the most efficient behavior of gathering their scattered food. I assumed further that, in order to enhance their encounter with food trapped in the surface film, they would also have to keep moving. And the faster they moved, the greater their chance of encountering something suitable to eat. But instead they were just idling. This meant that something interesting was going on, and somewhere in their behavior lay a challenging puzzle.

To figure out what the beetles were up to, more clues were needed. I would put a twenty-four-hour watch on them, and F. Daniel Vogt, a student in the ecology course I was teaching (who later came to Burlington to do post-doctoral work with me), agreed to help. We loaded our canoe with blankets, cookies, sandwiches, and a thermos of hot coffee, and then paddled out to the edge of one of the beetle aggregation areas. There were dozens of rafts, many of them containing hundreds of individual beetles. The rafts were along the edge of the lake and on patches of open water inside the boundary of the rushes. There were probably over 10,000 beetles in the total aggregation that stretched along thirty yards of shoreline.

We anchored ourselves in the rushes and waited. Nothing much was happening. The sun sank to the horizon, and the loons started calling back and forth across the lake. Blue herons flew overhead, silhouetted against a reddening sky. Still the beetles did nothing — it could be a boring project. But as it got darker, the beetles started to move, and about twenty minutes after sunset the beetles were milling about so fast in their rafts that the water looked ruffled. Periods of rapid, seemingly random movement by the aggregated thousands in any raft were interspersed with periods when the raft was still. Each raft acted independently, but individual beetles were obviously influenced by the others in their group. The beetles' tranquil periods became shorter, and a half hour after sunset there was no longer any pause in their activity. Also, the beetles were not holding together as tightly as before and their rafts were expanding.

Raft boundaries were becoming less discrete. Whatever force it was that had been holding them together was weakening. Beetles from the dozens of rafts intermingled, and the whole aggregation area was now seething with beetles zig-zagging in all directions. Against the light it looked as if a giant pepper shaker had been emptied on the water and all the grains were skimming about, stopping, rotating. In addition, some beetles were leaving the area altogether, striking out in straight lines down the shoreline. As beetles sped on the surface of the glassy water against reflections of moonlight, they engraved it with broad V's that could be seen from a long distance. Although many beetles dispersed, most of them zipped about in the aggregation area.

In the daytime we could not get within ten feet of the beetles before they bolted away, but now they were skimming right up to the side of our canoe. They were in constant motion, as if searching for something, and they didn't notice us. We dropped the mosquitoes we'd been slapping onto the water, and the beetles grabbed them instantly. It seemed odd that the beetles could locate mosquitoes on the water surface in near darkness with ease, while failing to avoid us in the canoe.

Gyrinid beetles have four eyes, but it is doubtful that they are of great importance in detecting and capturing prey. One set of two eyes is

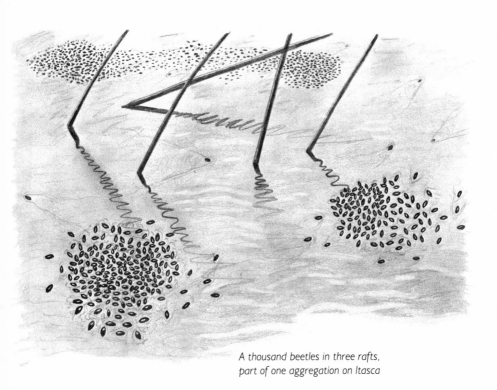

A thousand beetles in three rafts,
part of one aggregation on Itasca

directed up into the air; the other set is directed down into the water. Neither the air above the beetles nor the water below them contains the food they eat. All their food is on the water surface; eyes that are level with the surface should be of use only in seeing objects that stick up.

How then do the beetles find their food and orient themselves to their surroundings, especially in the dark? The question had already been asked in 1926 by Friedrich Eggers from the Zoological Institute at the University of Kiel. Eggers reported a set of observations of gyrinid beetles with surprising conclusions. He noticed that, when these beetles were swimming under water, they repeatedly bumped into transparent glass walls. However, when they were swimming on top of the water they easily avoided all obstacles, including glass, even when they were blinded by having their eyes painted over. He obtained the same results when observing the beetles in full darkness after marking them with lumines-cent paint. Eggers concluded that the beetles were not relying on sight to avoid obstacles.

He wondered if they were perhaps using their antennae. His detailed anatomical analysis of the beetles' antennae showed that the second segment at the base of the antenna is highly modified, containing an organ called the Johnston's organ that is adapted for detecting ex-tremely slight bending of the antenna at the base. The Johnston's organ is found in most insects, and it has evolved to serve a variety of functions. Since the base of the antenna rides on the water surface, and moves up and down with the smallest wave, Eggers speculated that beetles can detect small surface waves with their antennae. Further, beetles without the Johnston's organ were no longer able to avoid transparent obstacles. He concluded that the beetles avoid obstacles because of their ability to detect reflected waves that they themselves create when swimming.

Eggers' ideas could have been influenced by observations and discov-eries already made by others on bats and mosquitoes. It is impossible to sift out all of the factors leading to discovery. But a significant event that prompted our understanding of echolocation (used in this context as the detection of reflected waves) in bats, and beetles, was probably the sinking of the Titanic in 1912 after it struck an iceberg. Prompted by the disaster of the ship on its maiden voyage, a Maine handyman and engineer (and inventor of the machine gun), Hiram Maxim, thought about the problem of orientation without the aid of sight. In an article in Scientific American *in 1912 he expressed his conviction that bats could detect obstacles by perceiving the reflections of sounds caused by their wingbeats. With this inspiration he immediately thought of a practical technological application: ships could use the same principle by ringing bells and having devices to detect the echoes from icebergs and other obstacles. Within a year two patents were filed for devices to generate pulsed sound waves (in either air or water) and to detect the echoes from distant objects. The inroads that German submarines made in disrupting Allied shipping also stimulated development of echolocating devices. The bats and the beetles held the secret all along, though the generals weren't much interested in that unclassified information.*

Maxim's (erroneous) idea that bats could detect obstacles by perceiving the reflections of the sounds of their wingbeats undoubtedly was influenced by his earlier observations of mosquitoes. While fixing power-lines in Saratoga, New York, he had noticed large numbers of male mosquitoes that appeared to be attracted to humming electric lines, with which they attempted to mate. Maxim astutely observed the humming wires to have the same sound frequency as that of the female mosquito's flight tone. He thought that mosquitoes "hear" with their antennae, an idea that was rejected when he submitted it to the biological journals of his time, and so he wrote about it in a local newspaper. Maxim suggested that the plumose antennae of male mosquitoes intercepted the sound waves and were set in motion by them. After the antennae were vibrated by the sound waves, the movements were presumably perceived by the Johnston's organ at the base. Eggers' idea on orientation in the gyrinid beetles combined two of Maxim's ideas, since the beetles used waves for orientation (in this case capillary waves on water) and detected the waves by the Johnston's organ after the antennae were moved by those waves.

Maxim's seemingly bizarre ideas on mosquitoes have since been shown to be correct, but his more "reasonable" suggestion that the bats use the sounds of their wingbeats for orientation has not been substantiated. The idea was correct, but the details were wrong. In 1938 Donald R. Griffin, who was then an undergraduate at Harvard, determined that bats produce extremely high-pitched cries during flight that are inaudible to us. It is these cries, not the sounds of their wingbeats, that are used in echolocation.

Eggers did not test his hypothesis directly. Such testing might have required more technological expertise than was available in 1926. But a physicist, Peter Rudolph, at the Institut für Schwingungsforschung (Institute for Wave Mechanics) recently undertook this test. He recorded and analyzed the waves produced by swimming beetles and mimicked these waves with an electronic generator. He fixed live beetles in position and with a micromanipulator brought a thin glass rod to bear on the base of the antenna. The rod and the base of the antenna were moved at various frequencies and amplitudes to simulate the small reflected waves that the beetles would encounter on the water. Electrodes inserted into the base of the antennae wire-tapped the antennal nerve to determine if the beetles' nervous system could perceive the external stimuli. The nerve responded. Beetles therefore have the sensory capacity to detect the small deflected wavelets. The antennal nerve responded to changes in amplitude of only 0.5 μm and to frequencies up to 150 herz. At higher frequencies (which the animal would not likely experience) the response rapidly declined.

Since a fast-swimming beetle produces small capillary waves that extend more than six body lengths ahead of it, it might detect reflections of these waves produced off obstacles, prey, or other beetles from perhaps three body lengths ahead of it. Thus, a beetle covering 17 inches per second, can sweep one square yard for potential prey each minute. The beetle has no way of predetermining where food might be. It can only act to increase the odds of encountering food by moving far and fast when it is hungry.

After dropping a generous supply of dead mosquitoes down from our canoe, we were reasonably convinced that these beetles were nocturnal feeders, a conclusion contrary to reports in the published literature.

There would be much competition for insect prey in the dense aggregation we had seen during the day, and it should not take long before every insect on the surface film of the water would be found and taken. But it appeared that many of the beetles were leaving the aggregation at night to feed. If finding relatively stationary prey on the water surface is largely a matter of bumping into it, or coming close enough to detect it, then the more territory a beetle covers away from its competitors, the more likely it will find food.

As it got still darker we could see more of the beetles leaving the aggregations and swimming away rapidly along the shoreline just outside the weeds. We stationed our canoe about 25 yards offshore and some 275 yards down the shore away from the aggregation. There was a light on shore from the cook shack of the station, and its reflection cast a long bright band on the water. It was a calm night and the water had a glassy surface. When a beetle crossed the band of reflected light, we could see its V on the water surface from at least 25 yards away. We could, of course, also see in which direction the animal swam — whether away from the aggregation or toward it. At sunset there had not been a single beetle traveling along the shoreline. But after dark we counted 40–60 beetles every fifteen minutes during our watch. Before midnight those beetles we saw leave the aggregation were almost all going down the shoreline, away from the aggregation. From midnight until 4:00 a.m. they were traveling in equal numbers in both directions, and an hour before sunrise most of them were moving rapidly toward the aggregation. Nevertheless, the majority of the beetles did not leave the aggregation at all, but milled about close to the site throughout the night.

By the first light of dawn, the traffic of beetles away from the aggregation site was markedly reduced, and we resumed our observations at the site itself. There we saw single beetles approaching at high speed, and they appeared to stop in their forward motion when they bumped into the wakes of other moving beetles. When two beetles met they circled about, and then one followed the other. Still others came down the shoreline and were similarly stopped. Smaller groups joined larger groups, so that trains of half a dozen to dozens of beetles were formed, and these trains also joined the aggregation. The net rate of forward motion decreased as more and more beetles came together. One beetle by itself could swim in a relatively straight line and cover a lot of territory. But when it followed another, and that one reciprocated, both of their paths became erratic and forward motion was reduced. As more and more beetles came together, any one beetle alternately followed a succession of different beetles, and as a result none got much farther.

At daylight beetles ceased coming into the area that contained the rafts, but many thousands of beetles were now there milling and rippling the water surface. Soon they convened into many individual smaller and denser groups. Then their movements slowed down, and the rafts of relatively quiescent beetles were reformed.

The aggregation site we watched that night was not the only one on the lake's fourteen-mile shoreline. We paddled around the lake on numerous occasions and counted a total of only fourteen large aggregations composed predominantly of the whirligig species we were observing. (Many aggregations consisted of beetles of several species.) Each aggregation contained 50–100,000 beetles. Most of the aggregations were in the same location from one day to the next, while the size and location of individual rafts within an aggregation varied. But, during the two weeks in August when we made our surveys, we found that the size of some aggregations increased and that of others decreased. These observations were not consistent with the idea that the beetles always go home to their own group. Perhaps beetles that join a group at dawn are not necessarily the same ones that left it at dusk. In order to find out we needed to know what the individuals did, but first we had to be able to identify individuals by marking them.

To catch beetles for marking, we approached the rafts by canoe. I poised myself on the bow with insect net in hand, and Dan leaned into his paddle as we began our attack run. We achieved maximum speed at the rafts, and then I dipped with my net into the beetles as they tried to flee by scattering in all directions and diving beneath the surface. Still we managed to capture over a thousand beetles in less time than it took us to dab their backs with bright red paint, which also temporarily glued the wing covers together so that the insects could not fly. We released the painted beetles near the aggregation from which they had come, and the red beetles joined their black raft mates within seconds, where they quickly came to rest with the others. Late that afternoon we left satisfied with a paint job well done. The raft, with its many red beetles, was a beautiful sight. Because of the beetles' oily secretions, the red coats would soon be shed, but we hoped to see at least a few of our beetles again.

Some of the marked beetles would, of course, leave their aggregation at night. But would they return to it, or would we find red beetles in some of the other aggregations? We had an answer the next morning. Red beetles appeared in all of the aggregations in one arm of the lake. Some had traveled up to 2.5 miles. Clearly the beetles can join a group without homing, if they do any homing at all. Maybe beetles that strike out at night alone end up with others in the morning simply because they are stopped by others' swimming waves. Before dawn, still in the dark, they can follow one another, presumably using their ability to orient to capillary waves.

According to this hypothesis, a beetle that has been foraging most of the night need only cruise along the shoreline at dawn until it meets other beetles (which are likely to be close to a main concentration) and follows them in order to end up in a raft by daylight. We estimated that single beetles cruising along the shoreline had a net forward velocity of 84 feet per minute. Since the large concentrations were, on the average, a mile apart, it should take a beetle located halfway between two concentrations only 32 minutes to reach a raft. The large concentrations of beetles remained throughout the night in the same area, and so the aggregations were reformed there each dawn.

So far it is not known why some beetles leave the aggregations at night while others stay. A reasonable hypothesis (which does not necessarily make it correct) is that beetles leave to skim over the water surface when they are hungry and need to find food. Foraging success surely must be very low in the midst of an aggregation. The farther the beetle travels from the aggregation, the fewer competitors for the food.

Although it is not difficult to rationalize why the beetles might leave *aggregations, it is less obvious why they aggregate in the first place. I reviewed the different advantages for aggregation in other animals. Perhaps one of them might apply to these beetles. The situation is entirely different from that of social insects. In the latter many genetically related individuals may cooperate to overpower large prey (as ants do), or related individuals will communicate (ants, many bees, termites) to share resources that are lumped.*

Aggregations can also occur without cooperation. Many birds, such as crows, grackles, starlings, and finches, at times roost in large rookeries usually with others of their species, and individuals benefit by following each other to new foraging areas. On the water surface, however, the resources are randomly and thinly distributed. There are no good or poor foraging areas, and no beetle follows any other in **leaving** the group, only in returning to it. So for the beetles there can be few economic advantages to aggregating; to the contrary, it can only be economically costly.

Crowding into groups, though costly, can provide protection for some animals. For example, when an inexperienced hawk dives into a flock of starlings he is apt to become confused. If he treats the whole flock as a unit, then he usually emerges empty-clawed. Experienced hawks and other predators learn to focus on one specific individual when attacking; they ignore the flock or herd as such and doggedly pursue their target. But the members of the flock counter this strategy by crowding even more closely together when attacked. The beetle groups, on the other hand, are stationary, and when attacked they disperse. Beetles in the group can be captured by a fish without being pursued. Also, a fish could probably capture a half dozen or more beetles from one raft at a gulp, so it could treat the raft as a unit. A hawk can catch only one individual at a time; he can't grab randomly into the flock and expect to catch a bird.

The beetles were attracted to each other in spite of (or maybe because of) their bodily secretions. Water beetles exude noxious odiferous substances especially when they are touched or otherwise disturbed. Several German researchers became curious about these secretions because it was noted that some people in the Alps were able to elicit certain responses from their domestic cows and horses that had been fed some of these beetles, which suggested to them that the beetles possessed aphrodisiac substances. Working with water beetles closely related to the ones we examined in Minnesota, the European researchers found cortisone and the male sex hormone testosterone, as well as three other steroid chemicals, in the beetle's defensive secretions. The testosterone had the effect of narcotizing goldfish, making them float belly up. In addition, unsuspecting toads and frogs that gulped down these beetles violently regurgitated them in bloody mucus.

The beetles contain still other noxious chemicals in their secretions. In this country researchers have isolated another class of organic compounds — chemically identified as norsesquiterpenoid aldehydes, appropriately shortened to "gyrinidal," because they have been identified only from gyrinids. These substances are highly distasteful to a large range of potential predators. It seems, therefore, that the beetles have gone to some lengths, evolutionarily speaking, to make themselves noxious, at least to fishes, frogs, and toads. No wonder fish seldom seem to eat them, and perhaps their noxiousness may provide a key to their flocking behavior. There is only one thing that is more noxious than one beetle, and that is many.

If crowding together by these beetles is a strategy to amplify their noxiousness, then they are not unique. Most caterpillars that crowd together are noxious. Some kinds of stinkbugs form aggregations. In California, ladybird beetles when they stop feeding in the valleys spend the winter in large heaps up in the Sierras. Most of the noxious insects advertise themselves by bright colors. Gyrinids do not need to — they can be seen easily from beneath the water surface.

Noxious chemicals that make the insects taste bad may offer protection, but only against those predators who have tasted and likely killed them in the past. If the beetle's defense relies on the learned aversion of would-be predators, the chemicals can only work against those "educated" predators. We found, for example, that hungry fish in aquariums did not hesitate to snatch gyrinids conspicuously swimming on the water surface above them, provided they had not been previously introduced to the beetles. In the lake, young and inexperienced fish are being born all *the time, and the possibility that some fish forget their lesson after a while also cannot be totally discounted. Some beetles become casualties, and then others are protected. As Byron said, ''They never fail who die in a great cause.'' But no beetle wants to — each acts to survive.*

A beetle joining an aggregation can be quite certain that the fish in that particular area of the lake are already familiar with its kind; there have been victims before it. It thus faces very little risk of predation in that area. But a beetle that moves away from the flock cannot be certain that all of the fish it may encounter have been alerted to its noxiousness. The beetles try to stay together, like the great convoys of ships in World War II. The U-boat captains learned that the convoys were well protected. The beetles stay together when they can, not because they know of the danger they face when they are separated, but because their DNA has encoded appropriate responses, making them act as if they knew. Their DNA indicates ways of surviving in the hostile world on the water's surface. Over the last several decades we have begun to be curious about these secrets, but we have declassified very few.

Caterpillar Diners

While teaching ecology at the Lake Itasca field station up in northern Minnesota, I dreamed up and participated in some special projects. These were usually various sorts of tortures and delights, such as wading among hungry leeches out to an old beaver lodge to look at new plant colonization, or counting courting mink frogs at midnight near School-craft Island. Most of my students had not yet been weaned from their books. They were still nurtured by academic theories, which explained how things were supposed to be. The real world out there would have some surprises for them — from unexpected (and sometimes not repro-ducible) results to the persistent hordes of blood-sucking insects. Out in the field, in the real world, we would all find depth.

The patience of my group of six students was beginning to wear. For almost an hour now they had been peering at every leaf and twig on a small poplar sapling in front of them, looking for the three hidden caterpillars I had told them could be found there. Meanwhile, the mosquitoes and horseflies of Lake Itasca had been doing their job like no others. It had taken the students over a half hour to show the first signs of unrest. They were potentially good field biologists, or they would have resigned themselves more fully to slapping mosquitoes long before. It seemed inconceivable that caterpillars as long as someone's little finger could escape notice for so long, especially since the resolving power of the eyes was good enough to distinguish even the mites crawling on the bark. I had "planted" the caterpillars before the class arrived, and I could pick them out at a glance. Not because I had keener eyesight, but simply because I'd seen them before. I had a "search image" in my mind, and knowing what to look for made all the difference.

One of the students excitedly pointed out a mite as it crawled slowly up the stem almost on top of one of the caterpillars. If he could see a mite, how could he possibly miss three caterpillars, he wanted to know. I was able to persuade the six students to persist for an hour. But at the end of that hour three of the six still had not spotted all three caterpillars, though they had all seen the feeding damage that caterpillars had left on the leaves.

I originally became interested in caterpillars by wondering how they could manipulate and eat leaves of varying shapes. I had seen caterpillars bend leaves to eat the tips before eating the base of the leaf, where they had to remain attached. Bending the leaf, and eating from the tip, seemed to imply a specifically evolved feeding pattern, because it meant

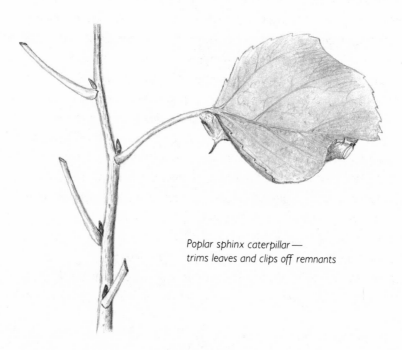

Poplar sphinx caterpillar —
trims leaves and clips off remnants

that by eating at the tip first the caterpillar could get to the rest of the blade later, thereby eating a whole leaf without wasting any of it. Eating from the tip of the leaf to the base, rather than vice versa, also meant that the caterpillar would not chew its perch out from under it. The caterpillars were not munching leaves in a random manner; they fed in systematic, adaptive patterns.

The caterpillar's behavior by no means implies the presence of forethought. The intelligence of caterpillars has already been irreparably downgraded by the famous nineteenth-century French naturalist, Jean Henri Fabre, who observed and commented on pine processionary caterpillars. These caterpillars occur in groups and, like our common tent caterpillars, have a tendency to follow the leader by tracking its silk trail. Fabre observed one caterpillar walking around a large vase, until it encountered its own track of silk and began to follow itself, like Winnie the Pooh and Piglet tracking the Woozle round and round a grove of trees, becoming more excited as they saw additional sets of presumed Woozle tracks with each successive circuit. Fabre's caterpillars became more and more entrapped by their track around the vase; the more tracks they created, the more followers they recruited. The group kept growing and kept walking around the vase for eight days, making 335 circuits. Except for some slight disorder that caused some to cast a few threads outside the circular path which eventually led them off, they would all have died on that vase.

The caterpillars naturally had not experienced a smooth vase in their evolutionary history. They specialized to feed on certain kinds of plants, and it was reasonable to suppose that they were programed with specific responses to make them do the "right" thing (if they are at the right time and place) without knowledge of consequences. But how were they specialized for different plants, physiologically and behaviorally?

Physiologically, caterpillars have evolved either to tolerate or to detoxify some of the strongest poisons that plants have evolved to defend themselves. As anyone who has seen a gypsy moth outbreak can testify, insects are supreme herbivores. And a gypsy moth is only one out of thousands of species that can cause great damage if not checked by disease and predators and by the plant's chemical defenses.

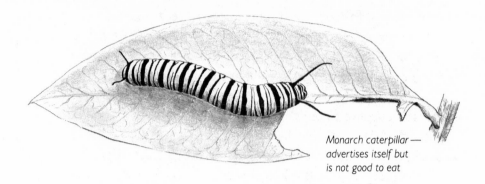

*Monarch caterpillar —
advertises itself but
is not good to eat*

A plant's poison may be effective against most herbivores, but there are almost always a few specialists that can breach the defenses. Indeed, some caterpillars even use the plants' poisons against them for their own survival. Perhaps the best-known example is milkweed and the monarch butterfly whose caterpillar feeds on it. Milkweed produces powerful poisons that affect the heart (cardiac glycosides) and taste extremely bitter. The monarch caterpillar, who is a specialized milkweed feeder, has a high tolerance for these glycosides. It even incorporates the poisons into its own tissues, and any predator trying to eat the caterpillars, or the adult butterflies, gets the bitter taste and soon learns to keep away. Few birds will eat a second monarch after having tried the first one, so both the caterpillar and the adult butterfly have evolved bright color patterns to advertise their identity to potential predators. It is difficult to miss one of the bright yellow-white-black caterpillars of a monarch sitting on a leaf. One researcher, Lincoln Brower, fed naive blue jays and toads some monarch butterflies reared on milkweed. The animals spit up the insects and would not eat them — or even butterflies that resembled the monarch — again. Brower succeeded in raising monarch caterpillars on cabbage, and these cabbage-fed insects became highly prized food. Obviously, it benefits the caterpillar to restrict itself to its proper food plant.

My point, though, is to suggest that the plant's strategy in defending itself has apparently backfired — at least against monarchs, who are one step ahead. If by someone's magic wand all milkweed stopped making poisons to protect themselves tomorrow, then all of the brightly colored monarch caterpillars would be sitting ducks to predators, and the plant might soon be rid of them. But of course this would then leave them open to other, cryptic herbivores. There is no such thing as being perfectly adapted. Evolution is a dynamic game, and of course there are already (still?) some clones of milkweed that do not produce poison. The players in the evolutionary chess game are always making moves, even though these moves are so slow that they are usually not perceptible in the short time frame of the flicker of a human life.

Aside from physiological responses needed to process or tolerate the tissue of different plants, caterpillars must also be behaviorally adapted to harvest leaves. The mechanical aspects of processing large amounts of leaf tissue are not trivial. Caterpillars cannot tear off whole leaves and chew them up whole as a cow can. They have to eat one bite at a time, and thus the shape of the leaf is of no less consequence than its nutritional value, since the animal must also perch on the leaf. Some leaves, such as those of basswood, are broad and flat, while others, such as those of the carrot family, are thin and composed of many fine thin branches. The caterpillar eating in small bites must manipulate the leaf so that it is consumed rapidly and with little waste, and so that the caterpillar does not lose its perch and fall to the ground.

The solution for handling leaves differs from one leaf shape to another. Different species of caterpillars utilize different plants, and their forms of leaf manipulation have evolved accordingly. In general, a small caterpillar can wander about on a leaf, feeding wherever the leaf tissue is soft and nutritious. A large caterpillar confronts a more complicated situation. In eating a large flat leaf it faces two problems. First, it must be anchored so that it is not shaken off; it cannot suspend itself from the thin blade and must remain firmly attached near or on the stiff leaf petiole. Second, if it simply eats all the leaf tissue that is near it at the petiole, then most of the leaf blade will remain out of reach and drop off uneaten.

The consequences of wasting food are particularly critical for larvae that feed on herbaceous plants with few leaves. The large larva of the sphinx moth, **Manduca sexta,** which can be a pest on our tomato and tobacco plants, provides an example. The caterpillar also feeds on a variety of solanaceous herbs that grow in the wild in isolated patches. Each patch of leaves is a limited source of food. The caterpillars feed almost continuously day and night. During the last larval instar (stage between molts), they grow from about 2 grams (the weight of a small hummingbird) up to 14 grams (the weight of a small mouse) in several days. During the feeding stage they can, if they find enough plants, consume over 678 square inches of leaf. Yet many eggs and larvae (of various ages and hence probably deposited by different individual moths) of M. sexta can sometimes be found on the same plant, whose foliage may be sufficient for only one caterpillar to complete development. After a plant is denuded, the caterpillars that have not grown to 10–14 grams must wander off in search of a new food plant, and there is always the very good chance that they will die before they reach a new food resource. Within limits, the faster a larva eats, the faster it grows. Their biting rate appears to be maximal at all times, regardless of how hungry they are, since caterpillars that have had continual access to food feed just as rapidly as those that have been without food for one or two days. Their leaf-chomping or biting rate (like a heart rate) is a direct function of their body temperature. At 25°C, for example, they take 2.5 bites per second. The higher the body temperature, the higher the leaf-chomping rate. (Any one larva is thus a living thermometer.) Eating constantly, except for brief breaks to digest, the larvae increase their body weight day and night at an average rate of 0.11 grams per hour, racing through the last larval instar (when they eat most of their food) in about four days.

On each plant, the different larvae are in an eating contest that is literally a matter of life and death. One might suppose, therefore, that there has been sufficient selective pressure for the larvae to have evolved the appropriate behavior to handle the leaves economically.

The tomato hornworm caterpillar, like most others, feeds by chewing off a series of leaf strips, much as we eat corn on the cob. The pattern of strips taken varies from one leaf shape to another. Many strips, parallel to the midrib, are usually eaten toward the leaf tip before the caterpillar encounters the midrib. It then chews partway through the midrib,

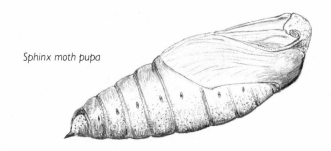

Sphinx moth pupa

greatly weakening it. The caterpillar continues to feed on the leaf ahead of its mouthparts, and the leaf bends of its own accord. The bending brings the tip area of the leaf within reach of the caterpillar, who meanwhile stays attached to the leaf petiole at the base. When half the leaf is pared down to the midrib and the caterpillar has reached the leaf tip, it begins to eat down the other half of the leaf by taking strips at right angles to the long axis of the leaf. Successive strips are taken perpendicular to the midrib, and this eventually results in the total consumption of the leaf as the caterpillar works its way back toward the petiole. The rule of feeding is simple: to eat food directly in front but to remain attached at the base.

The feeding pattern of these sphinx-moth caterpillars has evolved from the harvesting of flat leaves of a variety of shapes. But not all leaves are flat. A caterpillar could bite through thin pine needles in a few bites, for instance, but if it ate the nearest food first, it would harvest only a few mouthfuls before the rest of the needle dropped to the ground. How do caterpillars specializing on such leaves solve the mechanical problem of leaf manipulation?

I observed an elegant solution to this problem by the larva of another sphinx moth, the pine sphinx, **Lampara bombycoides**. This caterpillar remained attached to the pine twig near a tuft of needles. Each of the needles was several times longer than the caterpillar itself, but the caterpillar managed to eat the leaves in their entirety without moving from the spot. The caterpillar used its thoracic legs to "walk" to the tip of a needle, all the while remaining clamped to the twig with its posterior *clasper. In this way the leaf was bent like a bow, and the caterpillar deferred eating until it had bent the needle completely over. After reaching the tip, it ingested the leaf from tip to base by a rapid series of consecutive bites, as it walked the shortening needle back down.*

Feeding on anise and other umbellifers that are the favorite food of many swallowtail butterfly caterpillars presents a similar, though slightly more complicated, eating challenge. The feathery leaves of anise are not only thin and long but also highly branched. Any one leaf has dozens of tips, not just one. A swallowtail caterpillar that I observed feeding on anise behaved very much like the pine sphinx on short sections of leaf. This caterpillar also remained firmly anchored with its clasper while it continued to walk up the leaf with its thoracic legs. As the leaf was being bent, the caterpillar passed successive branchings of the leaf. It did not begin to feed until after passing the final branch point, when it finally reached one of the numerous tips of the leaf. When it had fed down one

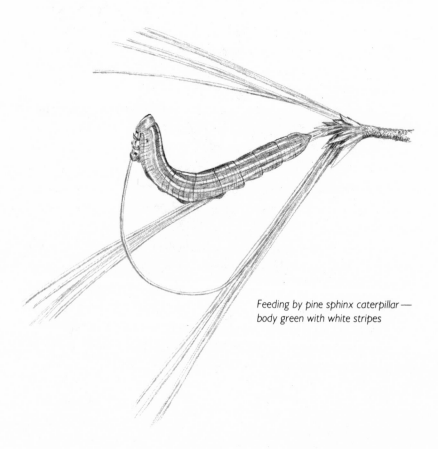

Feeding by pine sphinx caterpillar —
body green with white stripes

arm of a Y to contact the branch point, it momentarily stopped feeding and walked up the leaf (with its thoracic legs only) to the tip of the other arm of the Y before continuing again to feed from that tip. This simple procedure inevitably resulted in the total consumption of the leaf, and no leaf tissue dropped uneaten from the plant. These various patterns of leaf manipulation may seem obvious in retrospect, but to a caterpillar the solutions are hard-won over evolutionary time.

Biochemical plant defenses and leaf morphology may not be the only problems confronting a caterpillar when it feeds. While showing caterpillars to my students at Itasca, I observed some behaviors that seemed inconsistent with efficient leaf harvesting: something else still unexplained appeared to be going on.

One time I glanced at the ground and noticed a leaf from a basswood tree on the bare path leading to the dining hall. Undoubtedly I saw many leaves like this over the years, but I had never noticed them. There was something unusual about that leaf. For one thing, it was green. Trees don't normally shed green leaves in June. I picked the leaf up and saw that one side had been eaten off by a caterpillar. That was not particularly unusual either—what was strange was that this leaf's petiole was half as long as normal. Basswood petioles are very long, and petioles, the toughest part of any leaf, do not simply break in half. The petiole must have been chewed through. I felt like Sherlock Holmes. Did a caterpillar discard the partially eaten leaf? If so, I doubted that it was an accident.

The clipped leaf stood out as if flagged in red, because it didn't fit my expectations or theories of how I thought things ought to be. My immediate feeling was one of wonder. But the wonder was actually a composite of different theories that crowded my mind and vied with each other for validation or rejection. Did the caterpillar make a mistake? How did it evolve to afford that mistake? Was there something in or on the leaf that the caterpillar proximally, or in the ultimate evolutionary sense, was trying to get rid of? Did the caterpillar use the leaf as a parachute to ride down from the top of the tree to the ground in order to pupate? Different theories were buzzing around in my mind, and I would have to destroy *each of them until one, none, or a very few were left. Had I no theories at all, the partially eaten leaf on the ground would not have been noticed.*

Nothing right then could have been more exciting than sitting down in front of a caterpillar to watch it eat. I chose to watch a large brown animal that rests by pressing itself flat against the sides of basswood twigs. It was a **Catocola** *(the adult moths have bright red, orange, or yellow underwings). The caterpillar had markings and coloration that matched the gray bark of the branch almost perfectly. When sitting in place it looked like a slight bulge on the branch; it had apparently evolved to stay hidden from birds. All day the caterpillar sat without moving. In the evening it suddenly came to life. It turned around on the twig, crawled out onto one of the large leaves, and started to eat. It fed uninterruptedly for about an hour, eating less than half of the leaf. Then it backed down the petiole and chewed laboriously into the tough petiole. It was no accidental biting. In five minutes of hard chewing the leaf remnant dropped off to the ground, and the caterpillar promptly turned around and resumed its hiding place on the gray twig, persumably to digest until the next feeding bout.*

After this encounter with the **Catocola** *I observed the details of the feeding behavior of twenty other species. An interesting pattern emerged. Caterpillars from at least three different families (Noctuidae, Sphingidae, Saturniidae) chewed through the petiole near the base,*

Details of tip of abdomen
in a disturbed Abbott's
sphinx moth caterpillar

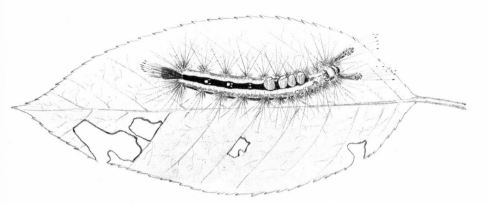

Tussock moth caterpillar — blue sides,
yellow stripes and tufts, red spot on head

discarding partially eaten leaves. This behavior, however, did not relate to their evolutionary pedigree. Other members of the same families did not do it, even on the same kinds of plants. Rather, the ones that chewed off partially eaten leaves were those species that were in some way cryptic in their appearance, either blending in with leaves or bark or, if visible mimicking something inedible, such as a bird dropping. All the leaf-clipping caterpillars fed on forest trees.

In contrast, none of the caterpillars that were spiny or gaudy clipped off leaves. These caterpillars ate only the choicest parts of the leaf, the soft tissues, leaving aside the hard and woody leaf skeleton that the clippers always consumed. The messy tatters and leaf skeletons left by the mourning cloak butterfly larvae, **Nymphalis antiopa,** on willow and poplar leaves, or those of the tough-skinned saturnid moth larvae, **Anisota rubicunda,** on maple leaves, or the colorful and spiny-haired tussock moth caterpillar, **Hemerocampa leucostigma,** and the arctiid moth larvae, **Halysidota maculata,** on maple, birch, alder, or cherry leaves — these were generally visible from a distance of several feet.

Any one behavior can and often does simultaneously have various functions, in the same way that many structures do. (The feathers of birds, for example, function for flight, insulation, sexual signaling, camouflage, to keep off biting flies, and possibly as light armor.) I would have to examine various potential functions. As a first approximation, it did not seem likely that the leaf-clipping behavior I had observed could be directly or primarily related to cutting off possible chemical defenses that the plant might mount to discourage the caterpillar's feeding. Leaf clipping occurred **after** the caterpillar had finished feeding and when the leaf no longer needed defending. Should caterpillars indeed attempt to avoid the toxic defenses of the plant, then all caterpillars should leave after a quick feeding. Birds who feed on caterpillars, though, are extremely keen in detecting movement; no matter how well camouflaged a caterpillar might be, if it moves its disguise is lost. One might predict, then, that they would evolve to become stationary. Quite the opposite is true. It is usually the caterpillars that birds do not like to eat who stay in place; the palatable ones who hide do the most traveling away from their feeding area.

Another explanation for why caterpillars cut off partially eaten leaves may be that they are getting rid of silk used to attach themselves with, since the silk can act as a scent marker to parasitic wasps and flies. But in that case one should not be able to observe the difference in behavior between species that are palatable to birds and those that are not. Both groups of species are heavily parasitized by insects, and so this hypothesis doesn't provide a complete answer.

Could predator avoidance be the key? Maybe a "clean" branch — one without damaged leaves — receives less thorough inspection by a hungry bird than one with leaves partially eaten by caterpillars. If birds zero in on leaf damage in searching for caterpillars, then caterpillars might disguise their feeding damage, get rid of it, or run away from it. I examined the behavior of more caterpillars to get clues. Most small caterpillars, such as the tiny geometrid larvae (inchworms), fed both in the daytime and at night. But after they fed on a leaf and left conspicuous damage, they would leave it to go somewhere else. Large geometrid larvae were active and fed only at night, but also moved away from partially consumed leaves; they remained rigid and immobile, resembling twigs, throughout the day.

Some caterpillars prized by birds (such as the notodontid larvae that mimic leaves or portions of leaves) did little traveling either day or night; nor did they clip partially eaten leaves. Their major commitment seemed to be to eat and grow. These caterpillars reduce apparent leaf damage by several mechanisms. First, they pare down the leaf evenly, gradually reducing its size but leaving its outline smooth. They never by-pass the tough and less nutritious portions of the leaf, such as the midribs, as the spiny unpalatable caterpillars do. They leave neither conspicuous holes nor tatters in the leaf, and it is not always apparent at a glance whether a leaf is partially eaten or merely of small size. In addition, most of these caterpillars eventually eat the entire leaf, leaving behind only the petiole. (After a leaf has been eaten, a tree generally abscises the petiole in several days.) The larvae feed forward on the leaf, bodily replacing the strip of leaf tissue they have just consumed, minimizing the apparent leaf damage by resembling the consumed leaf edge. Finally, they sometimes clip off the remaining portion of the leaf or the leaf petiole.

A caterpillar that eats flat leaves may disguise its past feeding damage, but can never entirely get rid of the evidence of its current meal. It may need one to two days to consume one leaf. Thus, even though caterpillars can use different behaviors to dissociate them from their feeding damage, it still pays off for birds to cue in on leaf damage. There very well could be continual selective pressure for the caterpillars to hide the evidence and for the birds to detect it.

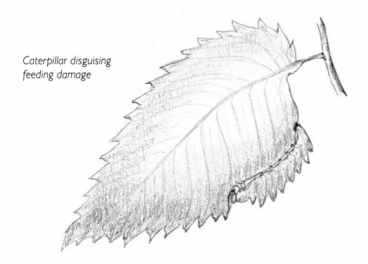

Caterpillar disguising
feeding damage

After my six students had searched an hour for the caterpillars planted on the sapling, I devised another experiment. I asked them to search in the forest for all the caterpillars they could find and to report on the methods used in their search. One student happened to find two larvae on a specific kind of bush and then looked for bushes of that kind to localize his search. Another discovered one in a rolled-up leaf and then searched for leaf rolls. But most formed search-images of the brightly colored unpalatable caterpillars and looked for them directly. All said they looked for leaf damage before confining their search to any one bush. The more leaf damage they saw, the more intently they searched.

Birds are also visual predators, and it is possible that they hunt in the same way. Individual birds have been observed to key in on leaf rolls, from which they unwrap the enclosed "hidden" caterpillars. Other birds characteristically specialize on one species of tree. Birds might also search for caterpillars using leaf damage as a cue. How can you test whether a bird sees leaf damage? You can't insert electrodes into certain neurons, show the animal the presumed stimulus of damaged leaves, and then watch for ways it might behave. This approach works for some simple animals and simple stimuli, but it would not work for such a

Chickadee observing leaf damage — caterpillar hides, suspending its head end by silk strings

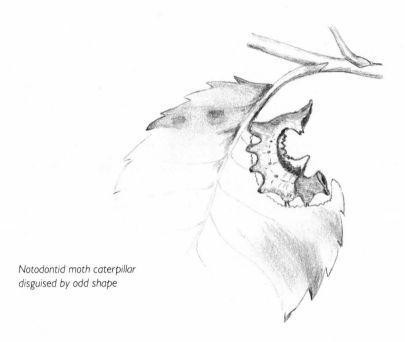

Notodontid moth caterpillar
disguised by odd shape

complex task. You would have to measure or quantify some kind of behavior under at least seminatural conditions. Ideally you would want fully natural conditions, but a person does not easily fly with a bird through the tree tops, to subject it there to tight experimental controls.

A large aviary might be a useful compromise, and several years later with Scott Collins (a colleague I had met at Itasca) I cleared a patch of forest at my camp in Maine and built a large enclosure out of a frame of felled trees covered with screening. Every day we put ten small new trees into it with fresh leaves, and on two of these trees we punched all of the many hundreds of leaves full of holes with a paper punch. (Later we used two trees with natural caterpillar leaf damage.) We removed all leaves with damage from the other eight trees. We loaded caterpillars onto the trees with damaged leaves, and then we let in a bird. We used black-capped chickadees captured in mist nets in the woods. We counted the number of hops our chickadees made in the different kinds of trees. After two months of observation we found out that chickadees can learn to search at specific kinds of trees and, furthermore, they can zero in on trees with leaves damaged by us or by caterpillars. When the prey was visible from a distance, such as mealworm larvae, some birds scanned for it directly. When the prey was very small or invisible, such as tiny cryptic caterpillars, the birds at first pecked into irregularities on the leaves or twigs, until (perhaps by chance) they grasped a caterpillar. Eventually they used leaf damage as a cue from a distance and specific search-image cues at close range. The sophistication demonstrated by the birds gave support to our hypothesis on the possible selective pressures affecting caterpillars' feeding behavior.

After many felled trees, some calluses, and a daily dose of blackfly bites, we had found out that the chickadees could use just about the same cues that graduate students do, but with more skill.

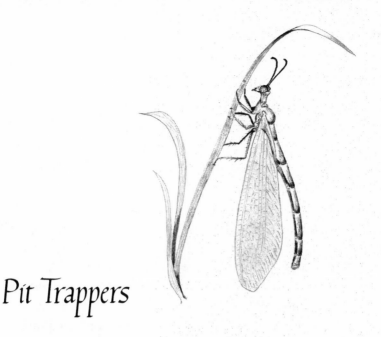

Pit Trappers

Algae-encrusted map turtles swam offshore with their heads now and then poking out of the water. They were waiting to haul in and bury their eggs in the sand. Many had already laid batches of eggs, and raccoons had dug up and eaten some of them. Leathery white eggshells were strewn about among small animal pug marks in the sand. Metallic-green tiger beetles heated by the sun ran rapidly on spindly legs, and flew up ahead of us when we came close to them. The larvae of these fast-running insect hunters are also predacious. They lurk in holes sunk into the sand, ready to snatch insects and spiders that come close.

I was delighted to be back in New England for good, having recently left Berkeley for a position at the University of Vermont in Burlington, where I was now exploring a secluded beach on Lake Champlain. Farther up the beach were dozens of tiny conical pits in the sand. I recognized them at once as the pit traps constructed by ant-lion larvae, which many children are familiar with and which I remembered from the Hahnheide more than thirty years before. I still found it irresistible to drop an ant onto the slippery side of a pit, to watch the sand being hurled by the ant-lion larva I knew to be buried in the sand at the bottom. Small sandslides ensued about the struggling ant. Eventually the ant slid to the bottom of the pit, and the larva grabbed it with its set of sharp hollow tongs that protruded from the sand. It immobilized the prey by dragging it under, and it would then inject poison, suck out the insect's juices, and fling the empty corpse back out of the pit.

Margaret and I stretched out in the warm sand on this sunny first day of May, admiring the perfectly conical pits and wondering about the behavior of these amazing animals. How were they able to construct such wonderful traps? How long would an ant lion have to wait before it caught something? Once a larva has made a pit, can it move and make a new pit in another location? How does it decide where to build? Is it always successful in catching anything that falls into its pit? Neither of us knew much about ant lions. We would learn by watching.

We first tried to get a general feel for what these animals were doing under natural conditions by observing them systematically. We marked out two adjacent plots of approximately one square yard each. The first contained 57 pits and the second 48. Then we reclined in the warm sand by the plots. Watching ant lions was more difficult than we had anticipated. After fifteen minutes or so propped up on elbows, I wanted to jump up, stretch, walk around. We set ourselves a time limit of an hour. The ant lions could wait interminably, but we could not. Man is also a hunter, but he is after all a more active hunter. He may have lived at the flanks of great animal herds, following them and preying on the weak, as wolves do today. The excitement I had felt hunting in my teens is only partially replaced now by the excitement of investigations. It seemed a particularly poor surrogate as we grew increasingly impatient. All the pits were uniformly round — the same — and my eyelids drooped in boredom. I wasn't **doing** anything.

We watched the plots over five one-hour shifts in five days. Eventually we did see prey captured, and when we did we cheered the patient little hunters. Some of the prey blundered in but then managed to escape, and we found ourselves feeling a slight disappointment. In order to get some idea why they were sometimes successful and sometimes not, we recorded the diameters of the pits, the size of the prey, and the response of the ant lions. We also examined the peripheries of other pits along the beach for corpses. Of 222 observed prey, only 36 percent were ants and 24 percent were spiders. We found a surprising number of winged insects (14 percent beetles, 12 percent midges, 9 percent small wasps).

Ant-lion larva in ambush

It seemed logical that the smaller pits were not as successful as larger pits in holding ants, and we tested the prey-holding capacity of pits directly by arranging for ants to walk over the lip of pits and slip down into them. We observed 126 encounters between ant-lion larvae having different pit sizes with small, medium, or large ants. All three species of ants that we used were common in the study area and were observed to be natural prey for the ant lions.

The smallest ants were easily captured by ant lions in pits of all sizes, although occasionally they escaped from the biggest pits, apparently undetected. The medium-sized ants, on the other hand, usually escaped from the small pits. Nevertheless, ant lions in small pits often responded vigorously to these ants by flicking sand up at the struggling ant. In one case a larva made 57 sand flicks before it finally captured the ant. The large ants had no trouble walking through the small pits, and the small larvae did not even bother to fling sand at them or attempt to capture them. All of the ant lions in medium-sized pits flicked sand vigorously at the large ants until the ant was caught or, more commonly, escaped. The large ant-lion larvae in large pits usually brought the large ants to their pincers within ten sand flicks. Clearly the largest pits were far more successful; they captured not only the smallest ants but also the largest, and with the least sand-flipping effort.

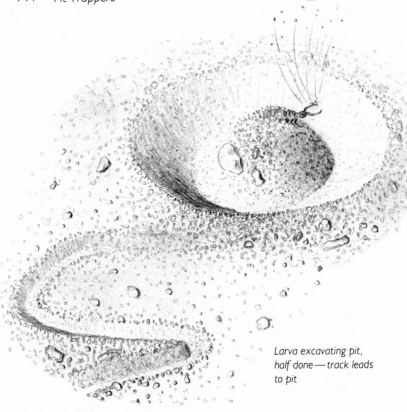

Larva excavating pit, half done — track leads to pit

From our observations we expected that the ant-lion larvae would dig the largest pits they could in order to enhance their prey-catching ability. We were therefore surprised to find a tremendous variability in pit sizes in similarly sized ant lions. We measured the diameter and depth of 200 pits and dug out the occupants and weighed them. We thought at first that the pit size might simply be a function of larval size. Indeed, there was a direct linear relationship between maximum pit size and ant-lion weight. These data told us how large larvae **could** dig pits as a function of body size. But many of the larvae, particularly the large ones, had pits less than half the maximum size for their body weight. Were these larvae less hungry? Had they recently moved and did not yet have time to enlarge their pits?

We were just getting into stride on the project on ant lions at the beach when we had to leave to spend the summer at our camp in Maine.

There we were expecting Dan Vogt, with whom I would do a field study on thermoregulation in dragonflies, and Scott Collins, with whom I would work on birds foraging for caterpillars. Almost as an afterthought we decided to take the ant lions with us. That was easier said than done. We captured one hundred ant lions easily enough. But they needed clean sand, and we had to take along some twenty pails of it because there was no beach sand at our camp. My wisdom in taking the ant lions to summer camp was much debated, especially as we were lugging the sand up the half mile of steep narrow trail through the woods.

It was the height of the blackfly season, and in my haste I slipped while chopping a tree, twisted my leg, and broke the medial meniscus of my right knee. Much of my scientific work in data collecting seems to involve physical activity. I'm somewhat like a hunter in pursuit of game, but no hunter ever chased a deer very far on a broken medial meniscus. People are adaptable, nevertheless, and if our "natural" pursuit is blocked, we usually follow another. I would now pursue a sit-and-wait strategy of hunting, like the ant lions themselves.

Confined to the cabin, I was glad we had brought the ant lions with us. I put one into a bucket of sand and sat back to watch. The larva backed up and flipped sand to the side and over its back, using its closed pincers like a shovel. After each sand flick the larva made another jerky motion backward, again covering its head and pincers with sand. As a result, it left a furrow in the sand as it traveled backward. In a few hours, furrowy tracks of the ant lion curved and crossed all over the bottom of the bucket. The tracks created in locomotion were, in effect, a long shallow pit in the form of a groove. The larva's mandibles protruded from the sand at the apex of the groove. I released several ants into the bucket with the larva. Small ants used the larva's tracks as walkways for several inches, and when an ant reached the ant-lion larva, it was frequently caught. (Long-legged ants, on the other hand, seldom followed the track and were not captured.) Clearly an ant lion did not need to have a perfect pit in order to catch prey. The tracks themselves aided in the capture. Pit building evolved gradually, most likely from larvae that buried themselves and waited for prey, occasionally flinging off the debris that covered their mandibles. If so, what behavioral steps were added to result in the making of deep pits?

There must have been some prior behavior that was seized on by evolution to be modified gradually into pit building. Was it the walking backward to hide in loose debris? The larva when it moved about in the surface layer of the sand often tended to etch roughly circular loops. I noticed that sometimes, if it made a relatively small circular loop, it got "caught" in its own track. If it then continued round and round in its own track, it eventually made a pit. When digging a pit, the larva made an initial loop of the right size and then simply moved in a series of smaller circles, flinging out sand, thus spiraling downward. The sand substrate itself determined the elegance of the pit trap. As the pit became deeper, less of the sand that was flipped up actually reached the outside of the pit. So, during the later stages of pit construction, the diameter of the pit tended to decrease slightly as the walls became steeper, and the larva moved in smaller spirals. The large sand grains were flung out of the pit, while smaller, lighter particles were retained on the walls. As more and more of the light sand adhered to the pit walls, the sides of the pit became ever-more unstable and subject to sanslides. In other words, the pit trap was set to "go off" at the slightest disturbance, such as an insect struggling.

The size of the initial circle defined the finished size of a new pit, but pits were usually enlarged later. Ant lions enlarged pits by backing partway up the wall and then digging in a ring around the wall. This undercut the walls above the ring, and gravity delivered sand from the edge of the pit, thus enlarging the top. The larva spiraled back down, flipping sand all the while, until it reached the apex of its new, larger pit.

If the ant-lion larvae are reflex automata, then one should be able to determine what their reflexes are by presenting them with the relevant stimuli. It must be kept in mind, though, that the animal's behavior has been programmed through evolution to make and use pits and capture prey—and that even relatively automatic behavior can appear to involve foresight or planning where none exists.

I dropped pebbles and other objects into the ant lion's pits and soon learned that the pit-trapping behavior is considerably more flexible and complex than it appears to be. The larva removed small pebbles by balancing them on its pincers and flinging them up and over the rim of the pit in one large heave. In contrast, large pebbles that it could not heave out were leaned on the tip of the abdomen and then gradually pushed up and over the rim as the larva backed up.

I once watched a larva weighing 44 milligrams trying to remove a 540-milligram clod of clay that had rolled into the bottom of its pit. It pushed the clod to near the top of the pit nine times, but each time after it had returned to the bottom and begun flipping out loose sand, the clod became dislodged and rolled back down. Finally, on the tenth try, the larva succeeded in pushing it completely out of the pit, and two and a half hours after the clod had first rolled in, the pit was fully repaired.

Objects that were too large to be tossed or pushed out of the pit were removed by a third technique. I dropped a wad of clay (1720 mg) into the completed pit of a 35-milligram larva to see if it could remove it. The larva dug around the clod, thereby collapsing some of the pit. It then backed over one side of the pit three times, lowering the rim in that area. It moved as far as two inches out of the pit, pushing back debris. When the wall had been flattened, the larva crawled back around the clod and pushed it out over the gentle slope of that side, then returned to repair the badly damaged pit. It is a complex automaton, indeed, that acts as if it understood about the laws of force and inclined planes.

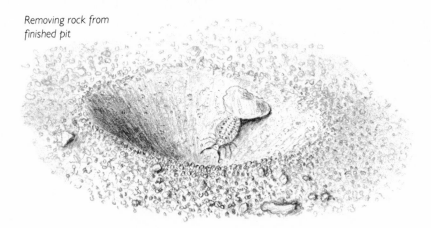

Removing rock from
finished pit

The time it took to construct a pit was highly variable and depended on the size of the pit, the temperature, the number and size of obstacles encountered, and other factors. At cabin temperature, a two-inch-wide pit was completed in ten minutes; another five-inch pit took three and a half hours.

Ant lions are also capable of chasing prey. I put ants into pits they had spontaneously abandoned. A larva, which had just been backing up over the rim, responded by turning a tight semicircle and backing down into its pit chasing the prey (and catching it). Ant-lion larvae normally did not enter constructed pits, and when placed into a "foreign" pit they immediately backed up and out. It is undoubtedly a prudent behavior, since ant lions will cannibalize their peers if given the opportunity.

Meanwhile our small cabin was becoming filled with buckets of sand, each holding two or three ant-lion larvae. We wanted to find out what they would do if they were hungry — would they enlarge their pits, move them, move them as well as enlarge them, or would they do nothing different at all? (We had to keep them in the cabin because outside the rain would have destroyed the pits every few days.) I had regained enough use of my leg to catch ants outside the cabin, and now some of the ant lions would be fed. For eighteen days I fed half of them one ant daily, and the other half were given no food; we measured pit diameters and mapped pit locations of both groups every day. We expected that the unfed larvae might build larger pits or move more often to build a pit elsewhere. But both groups of larvae enlarged their pits within the first three days of building, and then no further. Unfed larvae moved their pits occasionally, but so did the fed larvae. We didn't know why they moved at all, but our statistics showed that there was no significant difference in pit size or frequency of pit movement between fed and unfed larvae.

Eventually I tired of watching ant lions in buckets of sand and dumped the sand onto a level spot by the ledges near the cabin, where there were many ants. The larvae released onto the sand etched their furrows over it and then dug pits along the periphery. Larvae wandered through the sand until they encountered the edge of the patch. They followed the edge, eventually stopped, and dug. They stayed or moved at any one spot along the edge depending on sand temperature. At the hot beach in Vermont, where sand temperatures at noon soared to 134°F on the surface, the pits were built in the shade of peripheral vegetation. On the hill in Maine, where sand-surface temperatures rarely exceeded 104°F, the pits built in shade were abandoned and the larvae relocated to sunny portions of the plot where it was warmer and the sand dried out more quickly after the frequent rain showers. In one sand patch there were three ant-nest entrances in the middle, but the lions still built their pits on the outside perimeter of the plot where prey traffic was lowest.

The pits were enlarged throughout the summer as the larvae grew, and then the larvae burrowed into the sand to build cocoons out of silk from which the winged adults emerged in July and August. Those larvae not growing to adulthood that summer would overwinter and hunt again another year.

When we got back to the university, we were anxious to read about ant lions in the literature. To our surprise we found that a large number of papers had been published. My impression from the literature is that, in the eighteenth and nineteenth centuries, observations were made to discover the Creator's wisdom in His plan of the universe and all that is in it. Ant lions happened to supply a striking example to their first observers of God's cleverness in making these insects act like knowing creatures. The pendulum then swung the other way in the early twentieth century, and ant lions were thought to be no more than simple mechanical robots that could be explained by the laws of physics and chemistry.

Both approaches were wrong. The first was made through ignorance of evolution as an organizing principle, and the second disregarded the fact that any animal is a product of history since life began, this history being scripted on the genetic code. The interaction of sugars, phosphates, purines, and pyrimidines on DNA can be understood in terms of physics and chemistry. But what makes the animal is the code enscripted on the DNA, not the DNA itself. Laws of physics and chemistry provide no clues to biological history.

Modern researchers saw behavior of the ant-lion larvae in terms of "optimal" foraging. They examined their behavior in view of what they predicted it should be if evolution had designed them in the best possible way. Various and conflicting predictions of what they should be were given. Very often the evidence presented to back up the predictions was weak, and it was often biased by the predictions. It was theorized that the reason ant-lion larvae arranged themselves around the periphery of any one sand patch was to minimize competition. One researcher stated that the larvae moved their pits in response to hunger, while another said that they enlarged their pits instead. Most maintained that pit size was strictly a function of body size.

Our data did not support even one of the above ideas. We were especially curious as to why others had found behavioral responses to hunger when we had seen none. I decided to redo the experiment to see if there might have been something we overlooked. Were three weeks insufficient for eliciting a hunger response in ant-lion larvae? Were the boxes of sand too small?

The next summer we repeated the experiment, using much larger boxes of sand, putting only one larva into each box, and letting the experiment run all summer long, not just for eighteen days. Everyone in the lab had a chance to feed half the ant lions one ant a day. The other ant lions received no food. As in the year before, there was no detectable difference in the behavior of the two groups of larvae in the first three weeks. After a month, however, distinct behavioral differences between the two groups started to show. Those that received one ant a day stayed put, while those without food left their pit on the average of once every ten days to build another pit. After the larvae had been without food for two months they moved around even more, leaving longer trails before sinking a new pit, and the new pits they dug became smaller and smaller. It seemed as if they were making temporary test probes farther and farther away from where they had trapped unsuccessfully. Finding the trapping poor, they did not invest in making a big pit, but kept moving after giving it a short try. In the Hahnheide we had done the same.

We learned that even carefully collected results can be misleading if the underlying context of assumptions is wrong. I had misjudged the time frame. Each animal you study behaves according to a different schedule, and so does each phenomenon you want to study. To get relevant results, you have to be in tune with time frames — the ant lions had been more patient than we thought.

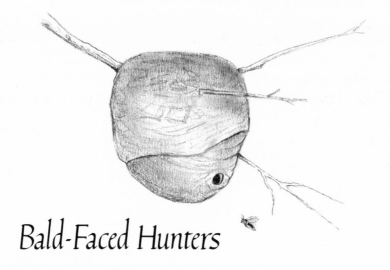

Bald-Faced Hunters

The football-sized hornet nest hung five feet off the ground in a red maple sapling about thirty feet from the edge of the dirt road. I cautioned my wife, Margaret, to stay in the car with the windows rolled up while I, encased in a bee veil and a heavy jacket and gloves, approached the nest with a stick attached to a generous length of twine. My plan was to come within throwing distance of the nest, to toss the stick into a fork of the sapling, and then to retreat into the car trailing the twine. I knew the wasps would go beserk at the slightest tap or touch on their tree because, when even one wasp is alarmed, it releases an alarm phero-mone (a scent) that alerts nestmates. But I planned to keep yanking on the twine, shaking the tree for a half hour or so, thinking the hornets would exhaust their supply of alarm pheromones and become peaceful. I could then get close enough to use my net to capture hornets leaving and returning to the nest. My purpose was to take their temperatures.

It didn't work out as planned. In the first toss the stick hit the sapling, but failed to lodge. In a twinkling the hornets came at me like speeding bullets, and I made a hasty retreat to the car. Five or ten minutes later I ventured out to try it a second time. By this time, though, the hornets were ready. They attacked before I even got close enough to throw the stick. Instead of having exhausted their alarm pheromones it seemed they had more! Eventually I managed to fix the stick in the sapling, but it dislodged after some yanking — long before the hornets had shown the least tendency to quit. I gave up on trying to exhaust their defenses. My theory was wrong. And my technique wasn't all that great either.

I then tried, without tricks, to catch individual hornets, but I think they now recognized their antagonist from a distance: stray hornets 35 feet from the nest were slamming into me and stinging. I wanted to get body temperatures of undisturbed hornets leaving and returning to the nest. But it soon became obvious that I couldn't get such data at this nest, since most of the hornets were attacking.

It was getting to be late in the afternoon, and I had only collected thoracic temperatures of attacking hornets, with no other prospects in sight. Stung quite a few times, I was becoming less amicably disposed. I finally decided to wait until dark to capture the whole colony in a bag, and then to kill the hornets so that I could inspect the inside of the nest in peace.

The first bats were flying and the stars were already out when at 8:30 p.m. the last foragers of the day were returning to my hornet nest. Under cover of darkness I approached and put an ether-soaked wad of cotton into the nest's entrance hole. Angry buzzing ensued. Then I stuck my thermometer, attached to Margaret's knitting needle, deep into the nest to measure nest temperature — 29°C and rising fast. More buzzing.

I worked quickly with bushcutters, snipping off the branches around the nest. The shaking caused the soggy cotton plug to drop out, and quite a few hornets came with it. They were alive and well, though groggy. I had to work faster now. In my haste I foolishly tore the nest envelope while trying to snip off a twig. More hornets came out, livelier ones. I finally managed (emphatically without Margaret's help) to dump the nest into a plastic garbage bag, along with half my bottle of ether.

The outside diameter of the nest, housing 152 adult worker hornets, was 8 inches. It contained three horizontal combs, each anchored by a stout pedicel to the one above it. The highest, and oldest, comb contained 65 eggs, 101 larvae, and 111 small (presumably worker) pupae. The second comb contained 25 eggs, 44 queen larvae, and 33 drone pupae. The queen larvae weighed up to 810 milligrams each, while the adult queens that these larvae develop into will weigh only half as much. There is, apparently, much shrinkage in the pupal stage. The nascent drones could be identified by their long antennae. The third and newest comb hanging suspended from the two above so far contained only 30 hexagonal brood cells, and each of them held one egg. Since the colony was nearing the end of its cycle (because it already contained nascent queens and drones), I suspected that these last eggs would develop either into queens or possibly into more drones. All of these nest contents were typical. What was unusual, however, was that the nest contained not one but two egg-laying queens with swollen abdomens. Two queens for one nest had never been reported for these hornets and similar wasps. One could dismiss it as "abnormal," but I found it an interesting observation. It shows what is within the realm of the possible in the real world. In biology there is no such thing as normal and abnormal, except in a statistical sense.

The whole brood, of 291 fat larvae and pupae, represented a sizable mass of tender protein. I donated it to the ant colony in back of our cabin. The ants appreciatively swarmed over the hornet brood, ultimately to convert it into more ants. That suited me just fine. This hornet nest was only one of many I molested over the summer. Still, molesting and all, I respected these insects. We had a long summer to get acquainted, and they became more and more interesting. I even got to like them.

I first noticed the white-faced hornets on our cabin door in early June. We awoke every morning to a scraping noise. It sounded like a mouse chewing on a dry cracker. But it was a queen paper wasp, the white-faced hornet, and she was chewing strips of gray wood fiber from the weathered outside surface of the door to make the first of six to eight paper envelopes for her nest. She paid no attention to me, even when I was close.

The queen made several trips every morning, and took more and more of my attention. After landing on the door she immediately scraped the wood with her strong mandibles. She took a thin quarter-inch swath and worked downward about an inch before she had a ball of wood pulp in *her mouth. Meanwhile, a light streak was left on the door where the wood fibers had been removed. The door was marked with many hundreds of streaks from top to bottom. Having gathered her wood pulp, the queen flew directly toward one end of the clearing I had made in the woods the year before. Somewhere in the piles of brush would be her nest. In early June the nest would be not much larger than a ping-pong ball, and it would be occupied by only one wasp, the fertile queen. By July the visits of the large wasp had stopped, and smaller versions of the same black bald-faced model appeared. These were workers, the sterile daughters of the queen(s) that had come earlier. The queen now stayed in the nest.*

The paper wasps have nearly the same life cycle as bumblebees. Each colony is started in the spring by a fertilized, overwintered queen. Throughout the summer the colony grows as the worker population builds up, and in the fall these sterile workers help to rear the reproducing offspring — the new queens (females) and the drones (males). (The workers, drones, and old queen all die in the fall.) The nest serves as an incubator to keep the young warm so that they will grow rapidly. The hornets have severe time constraints, since the summer is short. Their objective is to produce many reproductives at the end of the colony cycle in the fall, and this depends in large part on keeping the nest warm so that many workers can grow up rapidly. The cooler the nest, the more stunted the growth of the workers.

I speculated that the individual adult hornets must also keep their own body temperature warm in order to heat the nest and to be able to fly and forage for their larvae. No data on hornet body temperatures were available, and this explains my persistent behavior at the nest in the red maple. I wanted to find out if individual hornets maintain elevated and steady temperatures of the thorax, independent of air temperature. Wasps, unlike bumblebees, do not have fuzzy insulation covering their thorax. Would the hornets then be unable to elevate their body temperature? If so, would they be constrained in their activity at low air temperatures?

If thoracic temperature is not regulated during flight, then hornets leaving the nest should be close to nest temperature; those outside (returning to the nest) should have a thoracic temperature close to air temperature. All I had to do was stand by the nest with a net, capture hornets coming and going at different air temperatures, and probe each with the sharp point of the needle containing the thermocouple of my electronic thermometer. But we have seen that the hornets were uncooperative, and I found no magic means of calming them. They were usually peaceable enough while I made the first two or three measurements, if I approached quietly. But after that the alarm seemed to have sounded (or scented), and pandemonium broke loose.

There was another problem. I could not manage the delicate procedure of catching the hornets, inserting the thermocouple in just the right place in the second or two after capture, while wearing a suit of armor and heavy leather gloves. I needed lighter garb. There was no getting away from it—I had to sustain stings in order to obtain relevant data. Each day I collected a few thoracic temperatures, the number limited by my daily tolerable limit of stings, which didn't happen to exceed three.

At the end of the summer my graphs were filled with data, which showed that these hornets are superbly able to regulate their thoracic temperature. They are even better at it than bumblebees, who have the benefit of a coating of body hair. Furthermore, the attacking hornets were the hottest, with thoracic temperatures as high as 43.°C (108°F) at air temperatures close to freezing on a late August dawn.

Why were these insects apparently more warm-blooded than all the other insects that it might potentially eat? Because they are hunters? They catch other insects by pouncing on them before they take off in flight or by pursuing them in flight. A cold-blooded animal has slower reaction times. That is why I can easily swat the flies in our cabin early in the morning, when I have a high body temperature and they do not. However, when it warms up they evade me almost every time.

People are warm-blooded all the time. That makes it difficult to determine when warm-bloodedness is most advantageous or why it has evolved. We just accept it as a given. We cannot even tolerate a low body temperature, so we can't vary it to study it. Not so with these insects that can heat themselves up. They can tolerate a low body temperature, and they don't heat up unless there is an immediate reason for it. Afterwards they can cool down within a few seconds and then warm up again just as quickly. Small insects such as flies cannot heat up by their own metabolism at any time.

Hornets are living fly swatters. They live by catching winged insect prey. I wondered if they might take advantage of their ability to be warm-blooded at very low air temperatures when it would be easy to catch their sluggish prey. One way to start getting an answer was to ask when they get up in the morning.

Our alarm clock went off at 4:30 a.m. It was still dark, but the first hint of light could be seen on the eastern horizon. A robin started to sing, probably from the old apple tree in the clearing. Its clear flutelike tones drifted up through the pines to our cabin. Margaret and I walked stiff-legged out of the cabin, the cool fresh smell of morning in our nostrils. It was 5°C under a clear sky. It would probably warm up to 23°C by noontime.

As we walked across the field of spiraea bushes, our legs wet with dew, we passed a few stray bumblebees who had spent the night on the flowers. They were crawling feebly as we passed and lifted their middle legs in a defensive posture. They could not fly. Other insects sat here and there, even more immobilized. Suddenly a hornet flew by, pouncing back and forth onto leaves.

We hurried on to a nest I knew was hidden in certain bushes. I couldn't believe my eyes — hornets were leaving the nest and streaming back in. We started counting. There had never been so much activity at this nest later in the day, when I had observed it before. So the hornets were early risers. Later, after the sun was up and the other insects became active, their activity greatly declined. Whether or not the hornets were having better hunting success in the cool morning than in the afternoon is nothing I could document. I didn't have very good luck chasing individual hornets over the fields and through the woods to find out. A hornet on the hunt does not dally, as a bee does when it lands on flowers. It flies fast, swooping up and down, zig-zagging and bouncing into the vegetation.

A hornet on the hunt looks like a frenzied drunk. But there is method in its madness. Looking closer you observe that this bumping into almost everything is not accidental. Before each impact there is a fast sharp acceleration—the hornet is pouncing! Still closer examination reveals that most of the objects of the hornets' zeal are not what you might consider suitable food. One hornet, who I was able to follow for longer than most, pounced on twig stubs, spots on leaves, dried berries, a rabbit's dropping, and a piece of lichen. Finally, after what seemed like a long time, she (all workers are female) captured a small moth. Carrying the moth she landed on a twig, dangling head down by one hind leg, using her front legs to hold the moth. In seconds the head and legs had been chewed off, and then the severed wings fell like pieces of confetti. After another minute the remains of the moth were a macerated ball being worked rapidly by the hornet's mothparts into a mushy pulp. She grasped the ball in her mandibles, dropped from her perch on the twig, and flew straight to her nest at the other end of the field.

I watched other bald-faced hornets in the field and recorded 260 pounces. The hunters' pounces were, in order of frequency, onto the following items: leaves with brown spots (90), twig stubs (42), other white-faced hornets (18), yellowjacket wasps (18), flies (16), cracks on a log (13), bumblebees (10), seeds (9), spots on a log (6), discolored flowers (5), holes in a log (4), ants (4), bristly caterpillars (3), bumps on twigs (3), leaf dangling in spider web (3), raspberries (2), holes in leaf (2), bird dropping (2), solitary bees (2), blueberry (1), lichen on rock (1), wood chip (1), rabbit dropping (1), empty pupal case (1), siricid wasp (1), and the aforementioned moth. Besides that moth, the only other food out of the smorgasbord that I saw a hornet eat was a muscid fly. In summary, the wasps got 2 food items out of 260 pounces. Even though the hornet pounced every few seconds, only a tiny proportion of the pounces were on live prey.

The hornet did capture the bristly caterpillars, the ants, and most of the bumblebees on flowers, but they immediately dropped them. They captured less than half of the other wasps (other hornets as well as yellowjackets) they pursued and soon let them go, sometimes after a brief tussle during which the contestants were presumably getting more intimately acquainted. The hornets' policy seemed to be: attack first, discriminate later. They attacked objects that visually contrasted with the surroundings, whether in shape, color, or movement.

The hornets' environment is inhabited by hundreds of thousands of different species of potential prey. Most of these species are cryptic, resembling bark, twigs, spots on leaves, or sometimes even bird droppings. A hornet's life span is only a few weeks, a rather limited time in which to learn to become a good detective and taxonomist. In order to detect prey that have gone through millions of years of selective pressure to deceive predators, the hornets would have to be alert to the subtlest cues. So the hornet looks for contrasts, attacking almost anything, doing it very often, and making an occasional catch. As when taking a true-false test, in which there is little penalty for marking the wrong answers, it is best to hit everything. The only penalty the hornet pays for guessing wrong is that it loses time, although it minimizes this by barely slowing down in flight after each pounce. Once I had a very young pet owl that also pounced indiscriminately on knot holes, leaves, and tea bags. But it was learning by playing. The hornets weren't learning, though: their method of attack was simply the best way for them to find prey.

Some wasps scavenge from fruit and dead animals. Not the bald-faced hornets. A straight diet of insects involves a problem of energy — protein is not a suitable quick energy source for flight. Since wasps are perpetual-motion machines, what drives them? The likely answer is that they get considerable amounts of food energy from their own larvae. The workers feed protein to their larvae, which are amply stocked with digestive enzymes to convert it to sugar. They regurgitate the sugar to the adults, who accept it eagerly and feed the larvae more meat. Along with the sugar, the oral secretions offered to adults also contain metabolic waste products; the larvae's rectum is blocked, and there is no defecation until adulthood is reached. This arrangement assures that the larvae are well-fed, that there are no baby waste products smeared about the nursery, and that the workers are motivated by the sweet reward to attend to their young brothers and sisters.

The more I learned about the hornets, the better I liked them. Such order, underneath the surface blundering, is worthy of admiration, and I wanted to see what else they could do. So I investigated the paper they made.

A mature nest of the bald-faced hornets is enveloped in a half dozen to a dozen sheets of paper. One layer is attached to the other by small stout buttresses, so that the whole structure is sturdied. Each layer is laid down by the hornets using their mouthparts to shape the saliva-moistened wood pulp. A hornet draws out its ball of moist pulp into a narrow (3 millimeter) horizontal strip along an existing paper edge. Contiguous strips have slightly different soft hues of gray, brown, black, or dark green, since they are usually added by different individuals collecting wood pulp from different sources.

The paper must hold in heat, keep out water, and still "breathe." How is it constructed? The paper of any one layer in my nest seemed flimsy, thinner than the paper of the **Burlington Free Press**. I placed many layers together and measured the thickness with a rule. The newspaper had a thickness of 0.08 millimeters per sheet and weighed 4.7 milligrams per square centimeter. That of the hornets weighed slightly less (4.1 milligrams per square centimeter) and had a thickness of only 0.07 millimeters per sheet. However, the nest was composed of many layers separated by air spaces of several millimeters. The interior of the nest must be kept warm, approximately 29-31°C, and a paper envelope is essential for nest thermoregulation. Wasps kept in darkness at 32°C are known to build few or no paper envelopes, but if a nest is artificially cooled the wasps begin foraging for pulp and add more envelopes.

Holding my hornets' paper to the light, I could see that it was full of tiny perforations, like thousands of pinpricks. The holes allow for gas exchange and the escape of moisture from the nest's interior, much as the new fabric Gortex does. But they are not big enough for convective

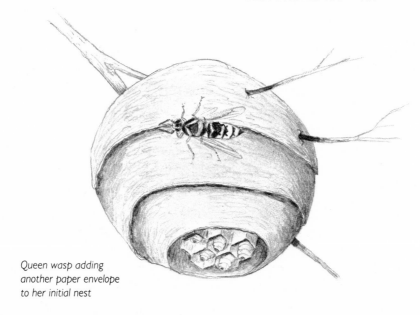

Queen wasp adding
another paper envelope
to her initial nest

air movements, and so the dead air from the layers traps heat and insulates the nest occupants while reducing the energy expenditure for heat production.

I subjected the wet paper to a stress test by taping a plastic bag to a strip 3 centimeters wide, attaching the strip to a support, and then adding water to the bag until the paper (the exposed strip having been soaked in water) broke in two. In three trials the most water the bag would hold before the paper ripped was 170 milliliters (170 grams, or 6 ounces). Thus, a 1 centimeter strip of wet hornet paper could support nearly 57 grams (2 ounces). I again compared this with the newsprint of the **Burlington Free Press.** Somewhat to my disappointment, the wasp paper came out of this test looking bad. The wet newsprint supported 250 grams per centimeter, leading me to conclude that the tensile strength of the newsprint was approximately 440 percent greater than that of the hornet paper.

I might have let it go at that. But then I looked at it from the hornets' perspective. The tensile strength of wet paper might be meaningless to hornets. Indeed, tensile strength is probably related to toughness, and toughness can be a liability. Wasps, as they expand their paper nests, have to chew up the innermost paper envelopes, and they then recycle the pulp by making new, larger layers on the outside. (This can be easily verified by adding dye to the outside of a small nest. As the nest grows larger through added layers on the outside, the color will first be covered by new paper, but eventually it will appear in the outermost layers of later construction when the nest interior is torn down.) The tougher the paper, the harder it would be for the wasps to chew and recycle, and the less it could breathe to let out the moisture produced by the wasps' metabolism.

How could such a flimsy material that I could see through withstand the constant rains of summer? Would the nest dissolve? A fairer test of the paper than a stress test, from the hornets' perspective, was to see if it shed water. I tore off a piece of the paper and placed it in a cup of water. The paper was not water-repellent and became soaked almost immediately. But what would happen if I drenched the nest? Determined to give it a severe challenge, I went to the spring and poured cupful after cupful over it, for ten minutes without pause. As expected, the top layer of paper immediately became wet. However, even after ten minutes, there was no sign of any dissolving of the structure. Indeed, though wet, the nest remained so strong that the paper did not collapse into the layer beneath it. After the paper was soaked — after the first cupful was poured over — the subsequently added water merely ran off over the top; for the most part, the nest was not even wet beyond the first layer.

When I got back to my laboratory in Burlington, I made use of a chronically dripping faucet. I turned the flow up just slightly and set a hornet's nest under the running water. One week later the inside of the nest was still dry. The paper might not measure up to excellence by our standards. But it is obviously good enough to do its job for the hornets. Evolution demands no more than that.

Counting Yellowjackets

I do not need numbers to tell me that the feathers of the scarlet tanager are red and stand out against the foliage. Unless I want to make a specific comparison, numbers will only intrude and detract from the bird's beauty. I do not need numbers to tell me that the sand wasp uses the tip of her abdomen to tamp down the soil she has scratched with her legs over the hole where she has buried her prey, a paralyzed spider. But not all things that are there to see are so unvaryingly clear-cut. I want to see what is out there in the field that does not meet the eye at a glance. I want to know not only what the wasp's paper looks like, smells like, and feels like. I also want to know some of its other qualities, besides those of surface appearance. Words like "thin," "strong," "fast," "large," "few," and "many" do not describe. They only reflect a lack of observation. How thin? How strong? How many?

The answers lie in numbers, in data. They distinguish between illusion and the sensible world of biology that I deal with every day. After having enough numbers, and analyzing them, one may be rewarded with a sudden "picture," or the dimensions of a portion of reality that was not and could not be seen before.

So here I was again, in the pre-dawn chill, counting the wasps leaving a nest and returning to it, writing the numbers in my notebook, and copying other numbers off the scale on a thermometer beside me, and *recording still others from a dial on my watch. I wanted to "see" how the wasps react to temperature. If someone had told me he had seen "some" leaving and entering the nest on a "cool morning" and "many" doing the same thing on "a warm afternoon," it would have been a waste of words. It would have said nothing. I needed numbers. My vision, the picture I would ultimately see, was totally dependent on them. The wasp this time was the common yellowjacket,* **Vespula vulgaris.** *Like the bald-faced hornet, it is a member of the family Vespidae. Most of the 15,000 wasp species of the world are solitary insects who use their sting to paralyze prey. Only the vespids are social, and they use their sting primarily for defense.*

The yellowjacket has almost the same life cycle as the bald-faced hornet. Each nest is initiated in the spring by an overwintering queen and on the average grows to several hundred workers by fall. In warm regions these wasps sometimes have perennial nests that grow to huge sizes. One V. vulgaris nest in California had 21 comb levels and was 4 feet tall. The adults that were taken from this nest (probably more than 100,000) filled four gallon jars. Another **vulgaris** nest in Germany had 21,692 cells for brood rearing, including 4,028 cells for new queens. One in Tasmania was reportedly 10 feet tall, with presumably tens of thousands of queen offspring. Since each daughter queen can potentially start a new nest, the potential for huge population build-ups of these wasps is tremendous. In some years, when the weather in the winter and early spring is appropriate, there are outbreaks in the summer and the wasps become severe pests. The fact that their venom contains histamine, serotonin, dopamine, noradrenaline, phospholipase A, phospholipase B, and hyaluronidase (all biologically active compounds in higher animals, including man), plus the fact that they show little hesitation in injecting their biochemical cocktail into any stranger who comes near the nest, has given them an unsavory reputation.

I was interested in them because they seemed similar to the bald-faced hornets, except in size. An average yellowjacket worker weighs only 50–60 milligrams, and most hornets weigh 130–180 milligrams. The small size of the yellowjackets might cause them difficulty in maintaining an elevated body temperature. If so, they might need a different hunting strategy from the bald-faced hornets.

Having found a nest of yellowjackets, I could grab wasps to take their temperatures as long as I dared. Like the bald-faced hornets, these wasps seemed to have an alarm system. After I had taken a few body temperatures, the wasps came out of their nest in the ground in clouds. But they were puny in size and not nearly so awesome as the hornets.

The thoracic temperatures of the yellowjackets were poorly regulated. Perhaps this might limit their activity. Indeed, I saw few yellowjackets entering and leaving their nests on the coldest observed morning, at 3°C, when there was frost on the ground. Numbers would allow me to see the wasps' relationship to the temperature of their environment, to visualize dimension of their wasp-ness, and possibly the beauty of their adaptations, which would otherwise not be possible.

After several days of counting wasps at various times and temperatures with the aid of Margaret and my graduate student assistant Jim Marden, I had a notebook full of numbers, but there was still no magic vision of things otherwise unseen. I plotted the numbers on a graph. Now I started to detect a pattern, but it might still be only an imaginary one. The spread of points that I saw seemed to say that common yellowjacket wasps were strongly affected by temperature in their activity, while bald-faced hornets were not. Appearance can be deceiving. But I could use the numbers to help give me an objective answer.

With a pocket calculator Margaret computed a regression line that best described the distribution of points. This line happened to fall close to what I had guessed by eye-balling the spread of points. For the bald-faced hornets the line going through the cloud of points looked nearly horizontal along the axis with 0°C air temperature at the left and 30°C at the right. For yellowjackets, on the other hand, the line rose steeply to the right. How much confidence could I give to the lines? Are the yellowjackets really affected by temperature, and the hornets not? I had many observations, many data points, and a simple statistical test then gave me the probability — 99 out of 100 — that the foraging activity of the yellowjackets is a function of temperature. Only now, through the help of numbers, I could see with a known degree of confidence what was not seen before. I could now plan my strategy to see through still another layer of opaqueness, until the wasps were exposed in ever-clearer definition. There is no magic in science. It is very often a slow, careful plodding.

The yellowjackets were more limited in their activity by their greater dependence on air temperature. Since they could not start out in very great numbers in the early morning, perhaps that precluded them from catching potential prey that would be fast and difficult to catch by the time they were up. They were, however, adept at supplying the protein to their young in the rapidly growing colonies. How did they do it?

I watched yellowjackets hunting in the field, and they seemed just as inept as the hornets. In addition, they landed on many leaves with spots that could hardly be mistaken for a fly or a cryptic caterpillar. Were they hunting by looking for indirect evidence, such as leaf damage left by a feeding insect? Numbers could help me to resolve this, too. I picked each of the hundreds of leaves I saw a wasp land on, and then with eyes closed I picked another leaf at random from the same bush. After having collected 170 leaf pairs I made a comparison. Twenty-two percent of the leaves chosen by the wasps had feeding damage, but 26 percent of the randomly chosen leaves also had the damage. The difference was not significant, and I concluded that the wasps were not seeing leaf damage or, if they were, they were not reacting to it. But I could still be wrong.

If insect meat is good, maybe other meat is even better. Our cat killed a woodland jumping mouse. I pulled off the skin and hung the carcass on a bush. Within minutes yellowjackets arrived, started pulling off strips of meat, and flying off with them. Within two hours the mouse was stripped. All that remained was a clean skeleton and the stomach contents. Two bald-faced hornets had also been attracted, but apparently not by the meat. They perched close to the meat, their heads jerking nervously every time a yellowjacket flew by. They often gave chase and sometimes even caught one, but always released it again; they didn't want the wasps. They also attempted to catch the blue and green blowflies that came to the meat to lay their eggs. But the flies' reactions were fast even at this time of the day. They almost always evaded capture. I observed only one of them captured and eaten out of twenty chases. The yellowjackets, meanwhile, seemed to be absorbed in taking the meat, ignoring the flies completely. Here, then, was a clear-cut difference in the foraging behavior in the two wasps: one is a hunter, the other is both a hunter and a scavenger.

If the yellowjackets specialize in finding and using large lumps of meat rather than catching small isolated prey, one might predict that they would communicate the distance and direction of such food to their nestmates, as social bees do. The next time the cat had a good hunting night and brought back a young rabbit and two deer mice, I divided the meat up into equal portions and distributed it at four stations in different compass directions around the wasp nest. Then we counted the number of wasps arriving at the stations as a function of time. Although more and more wasps came to each station as time went by, only a statistical analysis showed that the build-up of wasp numbers was linear with time. Again, the statistics made the difference between guessing and knowing. The conclusion was clear: the wasps do not communicate food location, even though they can communicate alarm and can mobilize nestmates to go out and search for food near the nest.

Why should yellowjackets so avidly take red meat while hornets ignore it, even though the meat protein is probably just as good as that from the

Yellowjacket

*great variety of insects hunted by the hornets? Why don't the yellow-jackets communicate food location? Undoubtedly there are some good answers to these questions. It might have been predicted that the bald-faced hornets take meat as the yellowjackets do. It might have been predicted that both, or at least the yellowjackets, would have evolved to hunt small vertebrate animals by stinging them to death. Perhaps they will, at some future time in evolution. If they were totally predictable by mathematical formulas, then there would be no need to watch them. As it is, they give me ideas of what they **might** do, and that makes it intriguing to see how well I know them, to keep watching them and asking new questions.*

Hunting Winter Moths

Almost Thanksgiving. The foliage has been off the trees for almost two months. It lies curled, crispy, and brown in a springy mat on the forest floor. Only a few stray oak and beech leaves remain rattling in the wind on the branches. The night frost has chilled the ground, and a light dusting of snow is beginning to press down the loose leaves. The sky is gray through the dark silhouette of maple branches. The snow smells good, but I can't describe it. Am I only observing the sudden absence of the musty-nutty smell of the fall leaves that I've been used to for so long?

My nephews, Chris and Charlie, are with me. They are carrying 30-30 Winchester rifles. (I'm armed with a butterfly net and an electronic thermometer.) Charlie is wondering whether to go to Bowdoin or the University of Maine next year, and whether to major in biology or something else. He works nights stocking grocery shelves to earn money for college. Chris, one year younger than Charlie, plays guitar in a local band. Can I tell them anything? Perhaps only that nature is fickle, opportunities change, and they should follow close on the heels of their interests.

We wear heavy jackets, insulated boots, gloves, and red stocking caps. Still, the cold is sucking out the calories. We had stopped near dawn at a diner to breakfast on eggs, sausage, home fries, and pancakes. It won't be long before we'll have an appetite again, but we are fortified with pocketfuls of oreo cookies and are prepared for both the white-tailed deer and the winter moths.

In Maine the autumn deer hunt is an institution as engrained and important culturally as the eating of the Thanksgiving turkey. To get your deer is a matter of pride, and in the Maine woods it is the first test of manhood. Also, one buck can provide a winter's meat supply for one person. Hunting is often the only way that a young person gets drawn out into the woods where he must be quiet, observe carefully and perhaps even think and wonder. This is how our ancestors learned about nature for millions of years. However, I don't justify hunting just because our ancestors have done it since the beginning of time, even before we became "human." No, I justify it because, if properly regulated, it promotes biological diversity. It is open to everyone, and it makes conservation a live issue. There are good hunters and bad hunters. But, in my observation, there are few people who care more about maintaining a healthy deer herd than hunters. They are often aware of what is there and not there, and they are willing to work to achieve healthy populations of grouse, deer, ducks, or bear. Mainly what you need is habitat, and when you have that you also have the thousands of other species that come with it.

As we continue to wander in the woods up onto the ridge, hoping to get a glimpse of a deer, we hear a hairy woodpecker hammering on the big maple tree beside us. A chickadee is calling from the forest. Then all is silent again. Snow is now falling lightly.

Many of the animals will escape the cold and the lack of food. They migrate or go underground. Under the litter of leaves and the first layer of accumulating snow, mice and shrews are already tunneling in a world of perpetual darkness, hunting for the millions of insects that are waiting out the winter in suspended animation, with antifreeze compounds in their blood. How, in this cold and the much greater cold to come, can the chickadees and the even tinier golden-crowned kinglets keep from freezing into blocks of ice, while we humans are suffering from cold feet already? It is a marvel that must be felt to be appreciated.

As we walk over the hardwood ridge, our footsteps making crunching sounds on the icy ground, I wonder how the small warm-blooded birds can find enough to eat to keep themselves warm and active. I peer into the forest through a lattice of gray maple trunks and see a flash of brown — a moth fluttering along, about four feet off the ground over the glistening snow, a small brown apparition. It seems so delicate, and the conditions so harsh, that it doesn't seem to belong here, but here it is.

Any warbler could catch it in one swoop. I capture it after a few bounds. It is the geometrid moth, **Operophtera bruceata,** that I had set out to find. The larvae of this moth are one of the many hundreds of inchworm caterpillars. The moth's antennae are feathery, as are the antennae of most male moths. The antennae are loaded with scent receptors to detect the female, who is wingless. The moth can probably smell only the females of its own species. **Operophtera's** mouthparts are atrophied. It does not feed as an adult, and the only food energy it has available is derived from the leaves the larva had fed on in the spring. It cannot get new energy resources — will it have enough to last until it finds a female to mate with?

I had seen these moths while deer hunting on the ridge before, but they fluttered past my consciousness as they fluttered past my sight. Charlie and Chris will most likely not get a deer today, but I hope they will notice the moths, and other interesting creatures of the woods, as they exercise their senses and imaginations looking for deer. Phil once told me that, when I got to be his age, I too would care less about shooting a deer than walking in the woods to see one. I am at that age now.

I have a specific reason for searching out the moths. All of the relevant publications on moths were on larger species that elevated their body temperatures close to that of our own, or higher. But I suspected that these winter moths might fly with a low body temperature, and I intended to examine this hunch.

To elevate their thoracic temperature for flight, moths are known to shiver before taking off, and that requires a considerable amount of food energy. I knew that the larger the animal, the more readily it can retain metabolically produced heat. We, for example, can maintain our body temperature near 37°C, even at subfreezing temperatures, provided we *have food energy. A shrew keeps the same body temperature as we do, but it has to eat several times its own body weight in food each day in order to do so. The sphinx moths I had studied in flight also maintain approximately the same body temperature as people and shrews, but they cannot elevate and maintain a high body temperature at air temperatures less than 15°C. Yet here are moths that are only 1/200th the size of a sphinx moth or a shrew, and they are flying when there is snow on the ground and air temperatures must be close to freezing. And they did not and could not feed. What adaptations make this possible? Are they flying with a low body temperature? It seemed likely that the moths could not heat up because they were so small, and they would have to be able to fly with a low muscle temperature unless a high muscle temperature is, in fact, an evolutionary necessity for muscle contractions at the very high rate required for flight. From all perspectives I could think of, it seemed a sure thing that these moths would fly with a low body temperature, but of course I had to measure temperature nevertheless. In biology, with its incredible complexity, little can be taken for granted.*

Air temperature was 3°C below freezing. I probed a moth with my thermocouple, and the instrument said that its body temperature was two degrees below freezing. Could that be right? A moth, flying at subzero temperatures, with a subzero muscle temperature!

"Hey, Charlie, Chris—keep your eyes peeled for moths!"

The boys now realized that the moths were also exciting game, and they adjusted their search accordingly. They forgot about deer. By the end of the day we had caught many moths. I was convinced by the data that, unlike many others, these particular moths, which are small and have no cover of insulating scales, do not heat up in flight. They could fly nicely even with a muscle temperature so low that it would freeze pure water into solid ice. A high temperature wasn't necessary for muscle contraction, as such. Like other moths, they flew at the temperatures their muscles are generally subject to in their normal environment.

I stayed at our hunting camp in the woods to watch the moths for another week. On two nights I got up at midnight and stumbled for about a mile with a flashlight through the underbrush. I made my way to the top ridge because I wanted to know how many moths might be active on the same transect I had walked with Charlie and Chris in daytime. The same area had to be sampled so that uniform comparisons could be made. The woods were very quiet. I heard a barred owl call once, and a far-distant crashing that quickly faded — possibly a deer, a moose, or a bear. Perhaps only a rabbit. The silence and my intent listening may have amplified the sound. There were few animals to be seen — but the moths were flying. Heavy predator pressure by summer birds and bats had probably led the moths to fly later and later into the season, until the primary constraint in their activity came to be temperature. If it was warm enough, they could fly both night and day, but on colder days they could only be active near noon when the sun was out. As the sun began to cast long shadows and temperatures dipped to −4°C or lower, they could not fly.

The adaptation of these moths is based on their uniqueness. They have survived by being different from the crowd. If all the moths of the habitat were to be active now, then this time would no longer be opportune because the predators would not have evolved to fly south. They would have stayed, to spend the winter feasting on moths.

What made it possible for these moths to be active in the cold? I noticed they have large wings relative to their small body size. Perhaps this has something to do with their ability to fly in the cold and to get by on their nonrenewable energy stores. Large wings may allow the moths to fly, in part, like a sail plane, with reduced reliance on internal power. This should make their flight less maneuverable but, then again, there are no bats and few birds to evade.

The idea that wing size is important was easy to test. My graduate assistant, Jim Marden, went around the campus of the University of Vermont at night picking moths off windows, to the amusement of the students within. I trimmed the borders of the moths' wings and then set these smaller-winged moths to fly in a temperature-controlled room. At any one temperature I trimmed off more and more wing until the moths could no longer fly. In this way I determined the minimum wing size for flight at high and low temperatures. When moths were flown at 10°C (50°F), for example, they could fly with about half their normal wing size. But to fly at −3°C, their minimum, they needed every bit of wing they had.

The effect of temperature on flight performance might also explain why the females had lost their wings. The flightless females are much heavier than the males because they carry a cargo of eggs. My graphs showed that a female, even if she did have wings, could not fly unless air temperatures were much higher than they usually are in the fall. Probably at one time in their evolutionary history the females did have wings, but they used them very infrequently, if at all, and the wings became useless. Wingless mutations finally arose in females, and these individuals used the food reserves otherwise used to produce wings to make more eggs instead.

The wing-clipping experiments gave me a further insight. They indicated at what temperature the moths could achieve the most power. The smaller the wings, the greater the power needed to stay airborne. Maximum power was achieved at air (and muscle) temperatures less than 20°C (68°F). At higher body temperatures, the moths are likely to suffer heat prostration. This suggests that their muscles are also uniquely adapted to function at low temperatures where most other moths' muscles contract slowly.

I do not have the expertise or facilities to do elaborate biochemical work. But one of the great advantages of having scientific contacts is that you can, with the aid of a telephone or a few postage stamps, tap into labs in most parts of the world. I sent samples of the moths to Thomas P. Mommsen, working at both Dalhousie University in Nova Scotia and the University of British Columbia. Tom has a special interest in enzyme adaptations to temperature. As I had guessed, he found that the enzymes of the moths' muscles that are involved in the breakdown of food molecules (stored fat) are extraordinarily active at unusually low temperatures.

The geometrid's solutions to the cold were not the only ones possible. Late one evening, as I was sitting under a large red maple, I noticed a moth of an entirely different kind. It was a typical owlet, or noctuid, moth. It was flying rapidly back and forth, inspecting the branches, and then it landed on the tree trunk close to me, continuing to vibrate its wings rapidly. It was shivering, contracting its flight muscles to keep them warm. I guessed that air temperatures had dropped to near freezing. I had never heard of or seen owlet moths fly at such low temperatures. This moth, unlike the geometrid, seemed to remain active at low air temperature not by tolerating a low body temperature, but by preventing body temperature from getting too low.

Winter moths were getting to be more interesting all the time, and I wanted to know as much as possible about them. I wrote and called several colleagues interested in moths and learned that the owlet moth was one of the Cucullinae, a small subfamily of the Noctuidae with 101 species in New England, active during the winter. Some may fly during any month of the winter provided that temperatures are not below freezing. But they fly only at night. They overwinter not as pupae or eggs, like most moths, but as adults, hiding under leaves under the blanketing layer of snow. Adults of both sexes fly. The moths mate and lay their eggs as tree buds break out in the spring, just before the migrant birds come back. They feed on the sap of injured trees and on

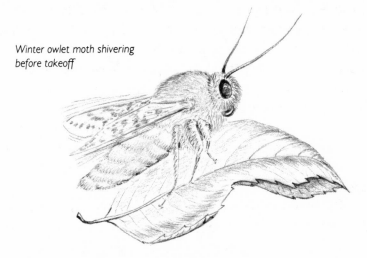

Winter owlet moth shivering
before takeoff

*red maple and willow flowers in early spring. They commonly fall into
sugaring buckets and drown there. Lepidopterists capture these moths
by painting trees with a thin solution of special concoctions of beer,
molasses, elderberry wine, brandy, rotten bananas, plum jam, cider,
brown sugar, oranges, or any one of a number of fermented variations
thereof. (Through systematic testing I later learned that the moths prefer
fresh diluted honey, and a solution of natural fresh maple sugar, to all of
the above-mentioned recipes.)*

*I couldn't wait to start catching these moths, and one day in early
March I went to a pizza parlor to get a large quantity of the first
ingredient — stale beer. I explained to the tap operator what I wanted it
for.*

"You want it for what?"

"To catch moths."

*"Yeah, sure, just a minute." He went in to talk with the manager, who
came out smiling, carrying a huge mayonnaise jar full of sudsy beer, on
which he had neatly written "Moth Lite."*

*That night I went out with Margaret and two of my students and we
applied our Moth Lite mixed with rotten bananas on the bark of trees.
The night was alive with moths. They settled onto the baited trees and
fluttered in the beams of our flashlights. We could pick off as many of
them as we wanted. The moths came in an amazing variety, all beautiful
creatures, with different geometrical markings and spots on background
colors of soft brown, gray, creamy yellow, orange, chalky white, umber,
caramel, and occasionally purple and green. It would probably take one
many years to be able to distinguish the different species, and I was
extremely grateful that there was at least one taxonomist, John Frankle-
mont at Cornell, who could identify them for me. For many groups of
insects there is no one in the world able to make identifications. There is
little point in writing a scientific paper on some insect's behavior or
physiology if one cannot assign the species name. Without the name no
one else can repeat the study or carry it on — it becomes a dead end.*

*Before I learned of the effectiveness of Moth Lite, I had written
Mamusha in Maine to ask her to save any moths that came to her sap*

buckets in February and March. She replied later that she had seen only two, and they had been frozen solid into the ice at the bottom of a bucket. She brought the block of sap ice into the kitchen to thaw out and, after the ice had melted, saw a moth fluttering against the window. Could the moths freeze into a block of ice and then fly away? It seemed unlikely, but I checked it out by freezing several of our moths in water in the refrigerator (set at −4°C). The next day I set the block of ice with moths inside it on a table. The ice melted, and the moths lay exposed. One of them stirred—a leg started twitching, then several. The moth righted itself, crawled a bit, and its wings began to quiver. It shivered for about five minutes and, just as it was lifting off for flight, I grabbed it and took its thoracic temperature. The moth had warmed up to 32°C (90°F).

I brought several hundred of the owlet moths into the laboratory and put them into the temperature-controlled room at −3°C. The cold made their muscles stiff. They could barely crawl, much less contract their flight muscles rapidly and strongly enough to fly as the geometrids had done at subfreezing temperatures. I turned off the lights and in fifteen minutes, in the diffuse light coming in through the door, I could see some of the nocturnal moths begin to come to life. They extended their antennae, raised themselves on their legs, and began to shiver, vibrating their wings. The frequency of the vibrations got faster and faster till the wings were a blur, and then they lifted off in flight. All of the moths that warmed up shivered to near 32°C before they would fly. Some of them could warm up and fly even at air temperatures of −3°C; but at this low air temperature few did so. Those that did could fly only for a few seconds at a time before the convection caused by air movements during flight cooled them off, and they had to stop and resume shivering to warm up again. At air temperatures of 4−10°C, these owlets could fly continuously, maintaining a thoracic temperature of 29−34°C. But at what to us is a comfortable room temperature (15−20°C), they were usually unable to remain in continuous flight for more than a full minute because of overheating. If forced to fly by being chased, they heated to near 38°C, and when they got this hot their flight became very weak and *they dropped to the ground in heat prostration. Adapted to fly at extremely low air temperatures, they have compromised their ability to be active at more "moderate" temperatures. The biochemical machinery of their flight muscles is adapted to operate near the temperature usually generated as a by-product of their flight activity. Since they are larger and better insulated, this temperature is considerably higher than that felt by the geometrid moths. But it is lower than that tolerated by sphinx moths that are active in the summer.*

The noctuid moths are endotherms (animals that elevate their body temperature by their metabolically produced heat). To some extent they also stabilize their thoracic temperatures, within a narrow range of low air temperature. Not only do these moths disobey the general insect rule of flying in the summer when it is warm, but they also do not conform to the generalities we see in other endothermic animals in relation to climate. According to Bergmann's rule, endothermic animals (birds and mammals) in the colder parts of their range are larger presumably because large size allows them to conserve more heat. Yet some of the largest endothermic moths occur near the equator, and the winter noctuids are tiny by comparison, only one thirtieth the weight of the largest moths.

The example of the different winter moths clearly shows the costs and benefits of different strategies of adaptation designed for the same end — to reproduce at low air temperatures when their potential major predators are normally not present. The geometrid moths succumb to the cold and have evolved to function at low body temperatures. The noctuid moths fight the cold, but have to expend more energy to keep their body temperatures elevated. As they say, there is more than one way to skin a cat.

Life on My Hill

There is a wild hill not many miles from our farm in Maine where Phil Potter used to take me deer hunting. In the fall the deer there turn the leaves searching for nuts on the beech ridges. Most of the beech trees are scarred by the claw marks of bear, while many of the young red maples have been stripped of some bark by moose in winter. I had always wanted to live in these woods, and once at seventeen I ran away from the Good Will School in Hinckley to do just that. I walked fifty miles through fields and forests toward the hill with my knapsack and my 22-bolt-action rifle, planning to build a cabin and live off the land. After walking two nights and a day I almost got there. But I didn't make it all the way until many years later.

Now Margaret and I are building a log cabin here, and I often walk through the woods to see the claw marks on the beeches, to follow fresh moose tracks in the snow, and to show these things to Erica, my young daughter, who has spent many days with me exploring clear down to the old beaver dams by our brook. My nephew Chris and I once played "otter," swimming downstream along the alders, clambering over dams, and poking under the banks to chase out trout. We were back last fall to chase winter moths. Now I know the hill well—certain trees where the pileated woodpecker, the barred owl, and the broadwing hawk have nested, sunlit paths where tiger beetles live, pools where wood frogs call in the spring. For me, this kind of familiarity breeds love.

Weasel in summer

My hill is not wilderness in the purist's sense. Yet it is returning surprisingly fast to what it once was. I remember remnants of old apple orchards within the forest where the deer and porcupine came to feed on November nights and days when the air was cold and crisp, and where the woodcock sang in the spring as the snow was melting. Most of the old apple orchards are gone now. Sugar maples and ash reach above the scraggly apple trees that now bear little fruit. Soon there will be a hardwood forest, with only old stone walls to remind us of the hardy pioneers. Deer and bear continue to feed on the few stray apples every fall, and I look forward to seeing their tracks, the broken branches and claw marks of bear on the ancient apple trees, and the rub marks of the bucks' antlers on the alders. I don't need to see these animals directly: anticipation is enough. I feel good knowing they are there at night. When signs are plentiful, I will hunt a buck; when they are few, I won't. Like any of nature's crops, there are only a certain number of apples. I would have them feed one bear, ten partridge, two porcupine, one moose, and one deer, rather than a dozen deer.

The tops of the ridges are still clothed in a dense growth of balsam firs and red spruces, and the swamp grass still grows tall among the alders down near the stream. Most of the abandoned farm, in between the ridges and the swamps, has been reclaimed by the forest. It is wild land again. In the woodlot north of an old stone wall, the trees now vary in age from very young to very old. Some are dead and decaying. Other animals will use, alter, and perhaps love this environment too. The hairy and downy woodpeckers will make holes, and after they leave the holes will be used by nuthatches and maybe bats and squirrels. The old spruces will harbor nests of yellow-rumped and blackburnian warblers and golden-crowned kinglets.

Instead of a farmhouse on the southerly slope, there is now a small clearing that is rapidly closing in from the edges. A cellar hole and a wellshaft, both reinforced by large stones, remain in the clearing. A patch of fireweed used by a ruby-throated hummingbird, hawk moths, and innumerable bumblebees grows inside the cellar hole, and at its edge are some thistles that the tortoise-shell butterflies feed on when they are in bloom and from which goldfinches harvest seeds in late summer and fall. An indigo bunting also sings in the spring from the now dead poplar that had grown out of the cellar hole. The bright blue male bird and its brown mate build their nest in the spiraea bushes close to the ground, as do the chestnut-sided and Maryland yellow-throated warblers who have also moved into the clearing. This is where we are building our log cabin, with spruce and fir from the surrounding woods.

Above the clearing on some ledges now sits a small hunting shack. When I bought this land in 1977 it was valued at $90 an acre, and the shack was thrown in free of charge. An old sign over the door read "Kamp Kaflunk." I go back often, spending four months there every summer.

The one-room hut sits on ledges beside several white birches. Among the rock ledges near the hut are several low open areas with acid soil where the rainwater collects. Here I have planted leatherleaf, wild rhododendron, lambkill, Labrador tea, blue irises, and other wildflowers liked by bumblebees and the flies who mimic them so closely that only a practiced eye can distinguish them. Each year the plants spread, and our investment and pleasure grow.

On the ledges is a small depression in the granite that fills with water. The yellow-rumped warblers and juncoes bathe there in the summer, and in May we usually hear a frog call from this tiny puddle. It is a male calling for a mate. There are always eggs in June, so apparently he is successful in calling a mate up to this high ridge. The tadpoles feed on the algae that grow from the nutrients brought in by the leaves that the wind drives to the water. There is not enough food for all the tadpoles. Many die, but their substance is passed on into that of the others, who finally hop out and wander away to catch unwary flies.

Once an ovenbird flew against the window pane of the hut and was killed. In the afternoon I dropped it onto the moss outside the door. Within hours a burying beetle came flying low over the ground. The black beetle with bright orange bands inspected the bird and then stood still with its abdomen held up into the air. It was emitting a scent to attract a mate. Soon there were two beetles, and before the next morning the bird was buried in the soil beneath the moss. In the meantime, before the beetle and bird were fully buried, a few blue-bottle flies had laid eggs on the bird. Within days maggots could have consumed the bird, leaving nothing for the beetles. But the beetles came laden with masses of tiny mites. These hitchhikers dislodged from their host while underground and would feast on the maggots, and thus the beetles were repaid. The beetles crawled round and round the bird, gradually dislodging all of its feathers. The bird became a globular mass of decaying meat, and into this the beetles deposited their eggs and then stayed to guard their young till they pupated nearby in the ground.

In the clearing I have enlarged around our camp, there is now a good cover of wild blueberry bushes. Another, slightly different habitat has been created. The hermit thrush, which sings in the evening from the spruces, sinks its nest cup here into the ground beneath a low blueberry shrub and lays four sky-blue eggs. Many insects use the area. I recently saw an **Ammophila** wasp with a geometrid caterpillar. It carried the caterpillar cradled under it and walked for twenty feet along the path to a patch of open soil at the edge of the ledge. It laid the caterpillar down, walked ahead a few inches, and lifted a few pebbles with its mandibles from a hole it had dug and camouflaged before. When the hole was open, the wasp returned to the caterpillar and dragged it into the hole. There it would lay an egg on the drugged and immobilized caterpillar, and the wasp larva would develop, eating out the caterpillar's insides. I also saw a small black wasp that seemed to watch the **Ammophila**. Would this wasp try to parasitize the **Ammophila** grub inside the caterpillar? Had the **Ammophila** evolved to camouflage its nest entrance to protect itself from parasitism? The wasp came back out of its hole, flew rapidly here and there, bringing back small pebbles and sticks, to recamouflage the nest entrance.

There are many other wasps. One I saw was a polished gun-barrel blue, with a pair of white spots on the abdomen. The wasp flicked her blue-black wings in apparent nervousness, as she faced a small round hole in the hard-packed earth. In front of the hole was a pile of damp,

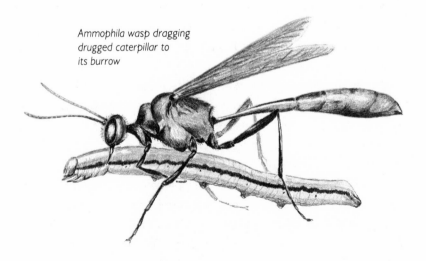

Ammophila wasp dragging
drugged caterpillar to
its burrow

recently excavated soil. The wasp backed down into the hole, abdomen first. Then she came back out, walked several inches to the dirt pile in front of the hole, stopped, and worked her middle legs into a blur as she kicked back a stream of soil over the hole. She backed up and tamped down the loose earth she had kicked over the hole, using the tip of her abdomen as a jackhammer. As the earth was tamped down, the hole again became visible and she came forward to kick more soil over it. She again tamped it down. Then she faced the hole and scattered the rest of the telltale pile of soil. She left, and I ran up to the cabin to get a pair of forceps to probe into the hole. It contained a huge orb-web spider, with a small transparent white egg attached to it. The wasp authority Howard E. Evans at Colorado State told me I had most likely observed the wasp **Episyron biguttatus**. The use of the tip of the abdomen in the manner I had observed is common in their genus. Some other wasps that bring prey for their young tamp the soil to close the nest entrance by using a genuine tool, a stone held in the mandibles.

Near the nest hole in the path, still another wasp that builds hard earthen pots for its young was collecting load after load of soil. I have so far not been successful in finding one of these earthen pots, and I look forward to seeing one. I will provide more sand and plant more wild-flowers, to attract more wasps. The wasps often scamper over amazingly hot sand. Are their legs long to keep their bodies from overheating near the blistering sand? How do they dig in soil heated to over 40°C—by stopping often to cool off? There might be interesting problems to work on here.

In late June many tiger swallowtail butterflies visit the red hawkweed blossoming in our clearing. They are only there for a few days before they lay their eggs and die. Later I see their green caterpillars on willow and pin cherry. Last year I saw a group of about thirty butterflies aggregated in one spot on the ground. I don't know what they were finding or how they were finding it. When I came back an hour later with my thermo-couple thermometer, they were all gone. Does one butterfly alight where it sees another sitting, in hope of finding something good to eat? Maybe next year I will pin down some dead butterflies to see if live ones are attracted. In some years it is quite hot in June. I wonder if the butterflies then sail more in flight than at low temperatures, to keep from overheating. Maybe sailing also reduces their energy expenditure when they have little to feed on. Bumblebees can't sail, so they must find rich energy resources. In the spring I often see them foraging at maple buds and on fir and spruce twigs. I wonder what they get there.

In the clearing there is already a population of little satyrs—small butterflies of the family Satyridae. The one that has colonized the clearing is the inornate ringlet (**Coenonympha tullia**); its caterpillars eat grass. It is a small pastel-colored butterfly, with predominantly russet-brown on the wings. It weighs only a quarter as much as the average unloaded honeybee. Its small size means it can't store much metabolic heat when it flies. Satyrids occur primarily in the New World tropics, and this one too is a lover of warmth. Just how much the activity of these little insects is at the mercy of temperature was not at first obvious to me. But then I logged twelve hours in the wake of these butterflies, following some individuals that remained in continuous flight—mostly to chase any other butterflies they could find—for over a half hour. I did not neglect to wear my running shoes or to carry a stopwatch, an electronic thermometer, an insect net, and a small tape recorder.

The butterflies showed a curious behavior. On hot days of 29°C (85°F) they flew nonstop for a half hour at a time, but on cool sunny days of 20°C (68°F), when I felt comfortable and could have chased them all day long, they elected to spend on the average 96 percent of their time perched on the meadow grass. When they did venture up out of the grass for a flight, it was brief, always less than twelve seconds at a time. It all made sense, after I started to take their temperatures and observe their behavior more closely. The butterflies were totally unable to regulate their body temperature in flight, but they needed to have a muscle temperature of at least 30°C (86°F) before they were willing to fly. These butterflies never shivered. Their flight metabolism produced enough heat to warm them up a few degrees, which is why they would fly continuously at 29°C. But the metabolic heat was not sufficient to heat them to 30°C at air temperatures of 20°C, which is why they didn't make long flights then.

When the butterflies were perched in the grass on (for them) a cold sunny day of 20°C – 24°C they assumed a very specific posture, apparently designed to maximize solar heating. They closed their wings and tilted themselves over on their sides to get perpendicular to the sun's rays. With this posture they could heat themselves up to about 43°C (109°F) in a minute or two. Once heated, they flipped over with their wings shading the body, and within another minute they cooled about 10° or they took a brief flight. In less than a minute of flight they had cooled below 30°C, and again they stopped and basked.

While chopping fir, spruce, and pine trees to enlarge our clearing and to make logs for the cabin, I noticed some other insects. Minutes after I began to limb out the tree, cerambicid beetles with curved tapered antennae three times their body length were clambering over the log in front of me. I know their larvae well. They are the huge white grubs that chew through the wood, leaving galleries. When small they start just under the bark in the soft moist cambium of a dying tree. Then, as they grow larger and their mouthparts become hard and powerful, they chew galleries in the solid wood itself, leaving sawdust behind that collects in huge piles outside attacked logs. They do not attack living trees probably because the sticky pitch secreted by those trees repels them. But what attracted them almost instantly to a felled tree that was still in fresh foliage and full of juice? How did they know that the tree would be suitable for the larvae's growth within days? Did the tree exude a death smell? Was it the smell of the pitch that attracted them — the very stuff that defends the living tree against beetle larvae attacks?

As I trimmed the limbs off the trees I also noticed the arrivals, with a clatter of dark blue wings, of horn-tailed wasps (Siricidae). There were two varieties. Both were shiny metallic-blue, and their legs were decorated with either white or orange rings. The larvae of these wasps also burrow in logs. But unlike the beetles, who lay their eggs directly on the log, the wasps drill their stout, short ovipositor into the wood. They walk over the surface of a log, palpating it wildly with delicate antennae that become a blur. Suddenly they stop, lift up their abdomens, and start shoving the needlelike ovipositor vertically down into the log. Their antennae then become motionless and stiff and remain so for several minutes of apparent concentrated effort, during which they seem oblivious of all surroundings. I peeled the bark off the tree underneath a wasp and noted that the delicate ovipositor had not only penetrated the bark, but was embedded in the fresh green wood beneath it.

I stacked some of the logs for firewood. In the meantime the growing beetle larvae did their chewing and spewed sawdust behind them through exit holes to the outside. A single larva can make a lot of noise when it chews wood. When many larvae of different sizes are chewing in several logs, they provide an eerie concert on warm summer nights.

A few weeks after the logs were stacked, certain ichneumon wasps arrived. A number of coal-black ichneumons with red legs and long ovipositors (1.5 inches) hovered around the pile. They landed here and there on the logs, probing their slender whiplike ovipositors into the holes of the pile from which the beetle larvae were shedding their sawdust frass. They didn't seem to be very discriminating. They probed into any hole with frass coming out of it. Did they lay each egg and let the larva fend for itself in trying to find a host in the vicinity? Or did they have taste receptors at the tips of their ovipositors and probe until they tasted a beetle larva, only then depositing an egg?

A different ichneumon wasp (**Megarhyssa**) came along. This one parasitizes the larvae of the horn-tailed wasps that were also in the wood. The ovipositor is inside a long sheath. The combined ovipositor and sheath trailed like a thin black thread for more than 4 inches behind the slender wasp (under 1.5 inches long) as she flew in and landed on a log with her spindly legs. She walked up the log, rapidly palpating with her antennae, dragging her ovipositor behind. She turned around on the log, walked back, palpating some more. She turned and walked up again. Back and forth she went, her excursions becoming shorter. When she had narrowed down on a spot, she rose onto the tips of her long legs, lifted the ovipositor, curled it partially under her abdomen, and placed the remainder vertically onto the log. It took a half hour or more, but the wasp managed to penetrate the rock-hard wood with her hairlike ovipositor, which was held only partially stiff by the sheath surrounding it. The tip of this ovipositor presumably tasted a sawfly larvae, penetrated it, then the wasp squeezed (by some unknown mechanism) an egg down the 4-inch length of her hair-thin probe into the body of the unsuspecting grub. The egg will hatch into another grub, feeding on the juices and internal organs of the first. Somehow the grub must circumvent the host's immune defenses. And it will now, instead of developing into a sawfly, hatch an ichneumon wasp.

Do these beetle and sawfly larvae have any defenses against their parasites? It is still a complete mystery to us how the ichneumon wasp can locate its hosts so deep inside the wood. Is it by scent, by metabolic heat production? Do they detect larvae by the vibrations of their chewing? If so, how do they discriminate one kind of chewing grub from another? A beetle larva might be more difficult to find if it chews at the same time that another larva in the same log is chewing. Is there perhaps a rough synchrony in the chewing bouts of the larvae? I listen. It's possible. I file the observation in my mind as a future puzzle to work on, a side project that might be interesting. There will be time to ponder the beetles and wasps when the wood is stacked inside this winter and I'm warmed by the fire.

Questions always jump out at me when I'm watching animals. Slowly I get drawn into the puzzles. I didn't plan to watch ants. But I couldn't help doing that as I was catching them outside the cabin to feed my ant-lion larvae. No doubt there is some sort of order to discover in an ant hill. But my casual glances at any mound, where I couldn't distinguish one ant from another and where I could watch one ant for only a few seconds at a time, seemed to suggest pure chaos.

Then I saw something different. Ants were running back and forth in a wide swath along the glacier scours in the ledges next to the cabin. Those running north were red (**Formica subintegra**). They were carrying ant pupae, larvae, and black adults of a closely related species (**Formica fusca group**). Both were about the same size. One species was taking and carrying another out of a neighboring nest. Did the reds win a war, and were they now taking black slaves?

The reds held the blacks by their mandibles, and the blacks all had

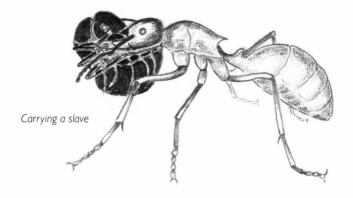

Carrying a slave

their legs retracted to the body, with their abdomen curled in under the chin of the carrying ants, apparently so that they could be transported more easily. Odd behavior, I thought, for a slave. Day after day for a week or more in June, I watched the ants as the nest mound next to our cabin was being emptied and the contents transferred to another mound (call it Mound A) 100 feet to the north. Later, in July, the red ants from Mound A took the contents of yet another nest. Slave-raiding ants are thought to take only unresisting larvae and pupae and let them grow to adulthood in their own colony, where the slaves then cooperate as colony members rather than resisting as aliens. These black ants were not resisting. Perhaps I was seeing a mixed-species colony with already integrated slaves (rather than a slaving raid); the colony could be moving from one to another of its possibly many nests. Still, from what I had seen, I couldn't entirely reject the hypothesis that these were slave raids after all.

During the next summer the two colonies that had been raided were empty, while colony A was doing it again. In July the ants from Mound A emptied yet another nearby mound. And then, on August 5, they were carrying in the contents of still another mound that was located 250 feet to the north in another clearing. The 250-foot-long column extended over the moss and through the low shrubs and the stone wall of our clearing, under the red maple woods, and then again onto the ledges, past dead ant mounds, and then to Mound B.

Ants loaded with pupae or other ants required fifty seconds (at 22°C) to traverse five feet of open ledges; thus a round trip for any one ant should have taken nearly seven hours. It might have even taken longer than that, considering that the woods were cooler than the sun-exposed ledges and the ants, whose walking speed is a function of their body temperature, would have slowed down. Throughout the trip to Mound A, none of the blacks resisted. None moved in any perceptible way. When I forcibly took a black from a red, though, it ran off rapidly — so the blacks were not physically incapacitated.

The number of black ants being transported was staggering. On one afternoon, with typical traffic between the two mounds, 63 percent of the red ants returning to the home mound were carriers. Of 138 carriers passing in ten minutes, 4 percent carried larvae, 62 percent carried pupae, and 34 percent carried black adult workers. It continued much like this, for ten days, except for interruptions during rain and at night. If the ants were a mixed colony and simply moving from one of their homes to another, they were going to extraordinary efforts to do it. On the other hand, if it was a slave raid, one wouldn't expect the victims to be so cooperative.

I opened the red ants' nest for a closer look. Piles of white pupae were then exposed to the sun, and immediately **both** the reds and the blacks busied themselves transporting the pupae down into the dark recesses of the nest. As I reached into the soft porous soil of the nest, the ants attacked my hand and the attackers were primarily the blacks. I scooped out handfuls of nest material and counted the ants — out of a sample of 201 ants, only 11 percent were red. Strangely, the blacks (the slaves?) predominated in Mound A, the raiding mound, yet 99 percent in the raiding column emitting from Mound A were reds. Only a few black ants were in the column. Had I not seen the columns of almost exclusively red ants for many days, I would have thought from this view inside the nest that it was a colony of black ants, with only a few reds sprinkled in.

It is sometimes useful for ants to **allow** themselves to be carried by nestmates. Ants often relocate their colonies when their mound becomes unsuitable. During these nest emigrations the ants not only carry their brood (eggs, larvae, and pupae) out of their old mound, but they also carry each other, with the adults tucking themselves into the posture I had observed which makes carrying easier.

I induced a partial nest emigration from Mound A by erecting a screen tent over this mound that let in sunshine but prevented cooling by the breeze. Mound temperature soared to 48°C (118°F) on one sunny afternoon. Over the next few days, whenever temperatures outside the mound were suitable, a column of ants was busily relocating the contents of Mound A to a new mound that the blacks were building 20 feet away. Now I could witness what I **knew** to be a nest emigration. Ants carried each other in the typical posture I had seen in the raiding columns, but there was a major difference: of 284 carriers I counted, 70 percent were black. Blacks carried other blacks, reds, and brood. Reds carried not only blacks but other reds as well. Both blacks and reds carried brood. Blacks were not only making most of the trips to and from the new mound; they were also working faster. I timed fifteen ants of both species in their column. The reds ran (at 26°C) at 2 inches per second, but the blacks averaged 25 percent faster, and they were slowed down by 15 percent only when carrying another adult. I don't know why adult ants of any one colony carry other adults during nest emigrations, but I suspect that it saves energy for the colony.

I suspect further that, in the interaction of the two species observed earlier, blacks did not carry other blacks simply because they were not attempting to move to another mound of their own colony. Clearly, however, they allowed themselves to be carried. Perhaps they were merely **responding** to the normal picking-up and carrying behavior, which in this case was not instigated by sisters (all worker ants are female) but by the reds. Did they do this because they were deceived into thinking the reds were really sisters?

Ants are known to be duped by various kinds of insects who live in their nests. These insects mimic ants in their movements and morphology, and ants feed them as if they were nestmates. The communication between ants and their guests has been well explored for over sixty-five years. In general, the insect guest cracks the code of communication normally used only among the ants' colony members. For example, the larvae of certain beetles who live with the **Formica sanguinea** in Europe show a characteristic begging behavior to their ant hosts, which is very similar to the begging of the ant larvae themselves. The ants respond to the beetle larvae as they do to their own larvae — by regurgitating food. The adult ants even groom the beetle larvae (apparently in response to chemical signals), even though the beetle larvae also eat the ant larvae.

If a beetle can successfully "impersonate" an ant to get food, it seemed possible that one ant could impersonate another of a closely related species. Why shouldn't the red ants exploit the blacks' weak spot (the response that allowed them to be picked up and carried) and use it to their advantage in order to take slaves? This would necessitate few modifications of an already existing behavior. They had already evolved to carry their own nestmates, and their potential victims had already evolved the response to allow themselves to be carried.

If the reds could pass themselves off as "friendlies," as nestmates of the blacks, by some subtle behavioral cues (maybe by acting in a nonaggressive manner) they would save themselves considerable casualties from fighting. If they could bring it off, they would get into the alien colony without having to fight the adult defenders before taking their larvae and pupae. The impersonation, if successful, would have another advantage: it would immediately give them cooperative adult slaves.

I used to dream of exotic animals in far-off places, the steaming jungles of South America or the plains of East Africa. I still do. But with time I'm discovering more and more excitement at my own doorstep. Much of nature is subtle, and it is difficult to appreciate it if one is used to the grandiose. I doubt that I would have stopped to watch a mere beetle, a bird, or an ant if I had had a toy train to play with when I was young — a train that rumbled and tooted and sped on fast tracks at the touch of a button. I became attuned to spending hours watching a bird just to see what it brought back to the nest, getting pleasure from discovering the subtleties. It is the subtlety of a bird or a carabid or an ant, multiplied a few million times over, that makes the whole. If one is not attuned to the

Eastern kingbird

fact of the first subtlety, then all the rest can pass unnoticed also, just as one sees only the train with the loud whistle rumbling past.

Living on my hill has given me a store of observations and ideas that vary in detail from sightings, like those of the beetles and wasps, to impressions and to qualitative observations, like those of my ants. Any one of them could potentially be expanded into a full-blown research project. Yet the projects I have already done on bees, wasps, and moths show me projects that still need to be done in those areas; the more questions you answer, the more are revealed. "Completed" projects are often jumping-off places for others, and thus you maintain momentum in a certain direction. I will most likely continue to be concerned with insect foraging, energy economics, and the physiology of thermoregulation, slowly allowing the pieces to coalesce to form a larger context. To gain deeper perception of how nature operates, I want to look at a variety of different animals.

For me, insects are a logical choice, at least in retrospect. There is no other group of animals that rivals them in diversity. They are found everywhere, often in astounding abundance. They can be observed at close range. They are not demanding, and I feel no moral compunctions caging them and experimenting on them in the laboratory. Also, they face many of the same problems of survival that we do.

Although a number of factors might predispose me to specific projects, I am also open to unforeseen events. Throughout my life so far, I have never been able to predict all of what the next day will bring. Here on my hill there are interesting things all around me. I don't know beforehand what I will see, what ideas I will have, and what ideas and observations of other researchers will make me curious about things I might otherwise take for granted. What I do know is that there is enough here to occupy me a lifetime.